DNA–Protein Interactions

DNA–Protein Interactions

Andrew Travers

MRC Laboratory of Molecular Biology
Cambridge
UK

CHAPMAN & HALL
London · Glasgow · New York · Tokyo · Melbourne · Madras

Published by Chapman & Hall, 2–6 Boundary Row, London SE1 8HN

Chapman & Hall, 2–6 Boundary Row, London SE1 8HN, UK

Blackie Academic & Professional, Wester Cleddens Road, Bishopbriggs, Glasgow G64 2NZ, UK

Chapman & Hall Inc., 29 West 35th Street, New York, NY10001, USA

Chapman & Hall Japan, Thomson Publishing Japan, Hirakawacho Nemoto Building, 6F, 1-7-11 Hirakawa-cho, Chiyoda-ku, Tokyo 102, Japan

Chapman & Hall Australia, Thomas Nelson Australia, 102 Dodds Street, South Melbourne, Victoria 3205, Australia

Chapman & Hall India, R. Seshadri, 32 Second Main Road, CIT East, Madras 600 035, India

First edition 1993

© 1993 Andrew Travers

Typeset in 10/12 pt Palatino by Expo Holdings, Malaysia

Printed in Great Britain by St Edmundsbury Press, Bury St Edmunds, Suffolk.

ISBN 0 412 25990 7

A catalogue record for this book is available from the British Library

Library of Congress Cataloging-in-Publication data

Travers, A. A. (Andrew Arthur)
    DNA–Protein Interactions/Andrew Travers.
       p.   cm.
    Includes bibliographical references and index.
    ISBN 0–412–25990–7
    1. DNA–Protein Interactions.   I. Title.
  QP624.75.P74T7 1993
  574.87'3282—dc20   92-38467
                                                                        CIP

∞ Printed on permanent acid-free text paper, manufactured in accordance with the proposed
ANSI/NISO Z 39.48-199X and ANSI Z 39.48-1984

Contents

Preface

Our understanding of the mechanisms regulating gene expression, which determine the patterns of growth and development in all living organisms, ultimately involves the elucidation of the detailed and dynamic interactions of proteins with nucleic acids – both DNA and RNA. Until recently the commonly presented view of the DNA double helix – as visualized on the covers of many textbooks and journals – was as a monotonous static straight rod incapable in its own right of directing the processes necessary for the conservation and selective reading of genetic information. This view, although perhaps extreme, was reinforced by the necessary linearity of genetic maps.

The reality is that the biological functions of both DNA and RNA are dependent on complex, and sometimes transient, three-dimensional nucleoprotein structures in which genetically distant elements are brought into close spatial proximity. It is in such structures that the enzymatic manipulation of DNA in the essential biological processes as DNA replication, transcription and recombination are effected – the complexes are the mediators of the 'DNA transactions' of Hatch Echols. In these manipulations the DNA no longer acts as a passive partner to the proteins – rather its physicochemical properties determine the preferred direction of the manipulation, be it the bending of the double-helical axis or the separation of two strands of the double helix. In this scenario the proteins in the complex both add precision and keep the DNA molecule under tight control in keeping with the philosophical tenet that biological systems leave nothing to chance. Similar principles of course apply also to RNA–protein complexes where, for example, the catalytic activity of particular RNA species is modulated by interaction with proteins.

The aim of this book is to give an overview of the function of DNA–protein complexes particularly taking into account the role of the structural flexibility and heterogeneity of the DNA molecule itself. In a short description of this kind it is impossible to be comprehensive. Instead, I have tried to choose particular examples which illustrate the general principles involved. Some of the choices may well be eclectic but, I hope, otherwise informative. Inevitably there are parts of this book that will be rapidly overtaken by new discoveries – the pace of contemporary research is now such that there are always new and interesting results that would merit inclusion. However a line must be drawn and I hope that the principles, if not the details, will remain valid. For the errors that remain I take full responsibility.

Finally I would like to thank all those who have provided the spur for the completion of this book. There are all my colleagues at LMB who have been a constant source of stimulation and a rein to the more wayward ideas. I am particularly grateful to Horace Drew, whose constant espousal of a DNA - centric view of the world constituted the essential genesis of this account. Thank you too to the always encouraging and incredibly patient staff of Chapman & Hall – it has been a real pleasure to work with them. Lastly, I am especially indebted to my wife, who has constantly encouraged this enterprise and without whose support it would have assuredly never been finished. It is to her that this book is dedicated.

Andrew Travers
Cambridge

1

DNA structure

1.1 STRUCTURAL FEATURES OF DNA

The initial step in the expression of any gene is the selection of that gene from among the many thousands encoded in a typical DNA genome. This selection invariably requires the interaction of protein molecules with specific sequences or structures in the DNA itself. To understand the physical and chemical basis for these interactions we must first consider the structure of DNA and the ways in which this structure can change.

The classical view of the structure of DNA molecules is derived largely from the X-ray analysis of oriented DNA fibres. In such fibres the preferred configuration consists of two antiparallel sugar–phosphate backbones wrapped in a right-handed double helix. The attached bases on one strand are directed approximately towards the axis of this double helix and form hydrogen bonds with their complementary bases on the other. On the exterior surface the sugar–phosphate backbones are separated successively by two grooves, termed the 'major' and 'minor' grooves. These grooves are defined structurally by the orientation of the base pairs (Figure 1.1) such that the N7 atom of the purine ring and the C5 atom of the pyrimidine ring face out into the minor groove. It is these exposed atoms, and any attached chemical groups, which are directly accessible to interaction with external reagents, such as proteins, certain drugs, or reactive small molecules.

A second aspect of DNA structure is the relationship of one base pair to its neighbours. For this purpose we can simplify the structure and consider each base pair as a planar domino (Figure 1.2), although in actuality the two bases in an individual base pair are usually twisted relative to each other about the long axis of the base pair ('propeller twist'). The degree of propeller twist is an important determinant of the external character of the double helix. The helical arrangement requires that each base pair be rotated relative to its immediate neighbour. This

DNA structure

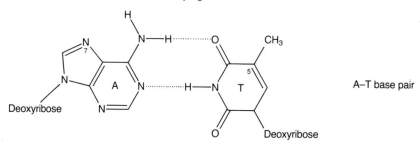

Figure 1.1 Geometry of A–T and G–C base pairs and their orientation relative to the major and minor grooves of DNA.

rotation is measured with respect to the local long axis of the double helix and is termed 'twist'. This twist is variable, but for most right-handed double helices it falls in the range of 22–45° between each successive base pair. A second important parameter is the angle between the planes of successive base pairs. The major observed departure from planar stacking is a rotation, termed 'roll', about the long axis of the base pairs. A rotation of this type results in a change of direction of the helical axis and thus can be directly related to the bending of a DNA molecule. A similar departure from planar stacking by a relative rotation about the short axis of the base pair, term 'tilt', is in general much smaller than roll.

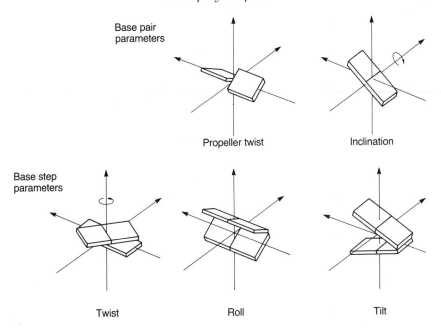

Base pair parameters

Propeller twist   Inclination

Base step parameters

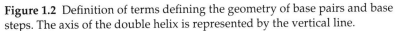

Twist   Roll   Tilt

Figure 1.2 Definition of terms defining the geometry of base pairs and base steps. The axis of the double helix is represented by the vertical line.

1.2 DNA POLYMORPHISM

The analysis of the X-ray diffraction patterns of DNA fibres distinguished two principal forms of DNA: B-DNA, which contains 10 base pairs (bp) per double helical turn, and A-DNA with 11 bp per double helical turn. These forms of DNA also differ in their internal geometry; in B-DNA the base pair planes are, on average, inclined at right angles to the double helical axis (that is, they have zero tilt), but in A-DNA the average tilt is +20°. The consequences of this difference are profound. In B-DNA the local helical axis is in approximately the same direction as the global double helical axis, but in A-DNA there is a roll angle of +20° between each successive base pair, which both opens the minor groove and means that the local helical axis follows a superhelical path around the global double helical axis (Figure 1.3). The different geometry is also reflected in the disposition of the base pairs relative to the DNA grooves. Whereas in B-DNA the base pairs are centrally placed in the double helix, in A-DNA they are displaced outwards towards the major groove, which is consequently narrowed and shallower relative to B-DNA. In addition, the rise for each double helical turn differs for these two forms; for B-DNA it is 34 Å, for A-DNA 27–28 Å.

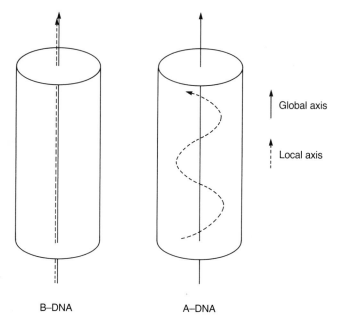

B–DNA   A–DNA

Figure 1.3 The relation of the local to the global helical axes for A-DNA and B-DNA.

   The stability of these distinct DNA structures is strongly dependent on the local environment and DNA sequence. The B to A transition is, in general, favoured by a lowering of the relative humidity. However, DNA molecules of certain sequences, notably poly(dA).(dT), fail to undergo this transition. A more important aspect of DNA polymorphism as seen in fibres is its relevance to biological systems. Short DNA molecules will crystallize from solution in a variety of conformations, some of which bear a close similarity to the classical B-DNA parameters and others of which are much closer in some aspects to A-form DNA. Nevertheless, even when such DNA oligomers crystallize in a form which resembles A-form DNA, the rise for each double helical turn is, on average, much closer to that for B-DNA. This suggests that, in solution, DNA molecules can occupy a spectrum of conformations which ranges from the classic B-form (as exemplified by A-T rich sequences) to forms which have some A-like characteristics but which are not truly A-form DNA. Such conformational variants are more properly termed A-type DNA.

1.3 CONFORMATIONAL VARIABILITY OF DNA

The ability of protein to discriminate between different regions of a DNA molecule depends first on the fact that the sequence of such a molecule is not uniform. This lack of uniformity is directly reflected in its local structure. Not only is the structure of the molecule heterogeneous but also the double helix is conformationally flexible and can assume different structures. The conformation assumed will depend both on the local sequence and on the local environment of the DNA. In general we can consider that a DNA molecule possesses two types of conformational flexibility: torsional flexibility, in which the twist angle either increases or decreases, corresponding respectively to a winding or unwinding of the double helix, and axial flexibility, in which the direction of the helical axis changes (Figure 1.4).

Changes in helical twist may be accompanied by alterations to the external dimensions of the double helix. As the twist decreases, so the minor groove becomes wider and shallower, while the major groove becomes narrower and deeper, although the sum of the major and minor groove widths is relatively invariant. Conversely, as the twist increases so the opposite changes are apparent. Some particular sequences have preferred twist angles, others are more catholic. For example, in solution, the homopolymer $(dA)_n.(dT)_n$ has an average twist of 36° per base step, equivalent to 10 bp per turn, and is thus overwound relative to the average twist of 34°. These same sequences typically possess a narrow minor groove. By contrast, DNA sequences highly enriched in G-C base

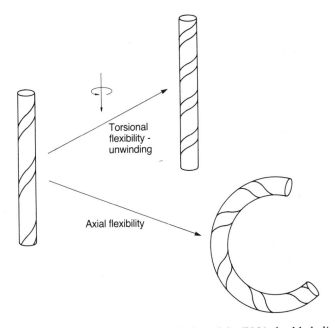

Torsional
flexibility -
unwinding

Axial flexibility

Figure 1.4 Axial and torsional flexibility of the DNA double helix.

pairs are somewhat underwound in solution and have a relatively wide minor groove. These fluctuations mean that the width of the minor groove will vary along the length of a DNA molecule in a reciprocal fashion to that of the major groove and this variation will thus constitute an external structural irregularity which can serve to distinguish differ-. ent regions of DNA.

The extent to which the twist of any individual DNA sequence can vary in solution depends on the base steps present in that sequence. For complete unwinding leading to eventual strand separation two factors are of major importance, the stacking forces between neighbouring base pairs and the hydrogen bonds between the individual bases in a single base pair. The stacking of one base pair on its neighbour is determined in large part by the distribution of charge within the base pairs. It is this charge distribution which imposes a preferred geometry for particular base steps. For example, in the homopolymeric sequence (dG).(dC), the charges on neighbouring guanine residues result in a repulsion which forces adjacent base pairs in a GG/CC base step to slide relative to each other. This consequently results in a displacement of the base pair from the double helical axis. Such a displacement resembles that found in A-DNA (Figure 1.5). By contrast, in the homopolymer (dA).(dT) neighbouring adenine residues stack with more overlap and consequently there is little or no displacement of the base pairs from the double helical axis as in B-DNA. Other conformations may be sterically determined. For example, for purine–pyrimidine and for pyrimidine–purine base steps the presence of purines on opposite strands in successive base pairs sterically restricts the conformations that these base pairs can adopt relative to each other. This effect, known as 'purine–purine' clash, is much more pronounced for pyrimidine–purine base steps than for purine–pyrimidine base steps. Such considerations suggest that the DNA sequences requiring the least energy for unwinding and strand separation are those which contain the fewest interbase hydrogen bonding interactions, that is A-T rich sequences, and whose stacking interactions are lowest, that is pyrimidine–purine base steps. The dinucleotide TpA satisfies both of these conditions and it is the base step that serves as a nucleus for DNA unwinding in many enzymatic reactions requiring strand separation. Similarly, the step TpG and its complement CpA are thermally less stable than most other base steps.

Again, for increasing the helical twist of DNA the interactions between neighbouring base pairs are crucial. Such an increase can be accommodated by a narrowing of the minor groove, a condition that is in general unfavourable for pyrimidine–purine base steps on account of the purine–purine clash. The sequences which have the lowest barrier to overwinding are A-T rich sequences. The selection of such sequences in the DNA may be favoured where biological function requires an increase in twist as in certain DNA–protein complexes.

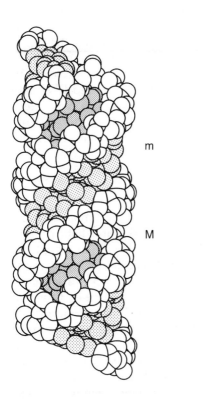

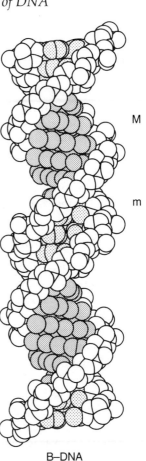

A–DNA

B–DNA

Figure 1.5 Axial projection of A-DNA and B-DNA.

The axial flexibility of a DNA molecule is another important factor in DNA–protein interactions. On average, double stranded DNA molecules up to 150 bp long behave in solution as stiff, but not necessarily straight, rods. However in many large DNA–protein complexes, such as those involved in DNA packaging or recombination, the DNA is often tightly bent. The archetypal example of such bending is the eukaryotic nucleosome in which 146 bp of DNA are wrapped nearly twice in a left-handed superhelix around a protein core containing eight molecules of histone proteins. In this complex the average radius of curvature of the DNA is about 43 Å, or just over twice the width of the DNA molecule itself. In a structure of this type the major and minor grooves are compressed on the inside of the curved molecule and stretched on the outside. At the same time the geometry of the base pairs must accommodate the change in the direction of the helical axis. Structurally these two requirements are closely related. A roll between successive base pairs results both in a change in direction of the local helical axis and a local change in groove width. For example, a positive roll in which the angle between the planes of successive base pairs opens out into the minor groove results in an expansion of the minor groove and a compression of the major groove. Base steps with this property will thus prefer to be placed where the minor groove is on the outside of a curved DNA molecule and the major groove is on the inside. Conversely, base steps which adopt a negative roll angle will compress the minor groove. Consequently, the preferred placement of such steps will be with the minor groove on the inside of the curve.

1.4 INTRINSIC BENDING OF DNA

In solution, a DNA molecule does not assume a single static structure. Instead there is a dynamic flux of changes in axial and helical parameters. This means that, in general, a DNA molecule will not adopt any preferred direction of bending in the absence of specific constraints, such as circularization or interaction with proteins, unless the structure of particular sequences is sufficiently stable and rigid to resist transient structural changes. The most notable example of stabilized bending in a DNA molecule, that is, a preferred direction of curvature intrinsic to the molecule itself, is characteristic of small circular DNA molecules found in the kinetoplast organelle of certain ciliates. This type of bending is detected by the anomalous mobility in polyacrylamide gel electrophoresis of short DNA fragments containing the bent region (Figure 1.6). The kinetoplast DNA moves more slowly through such a gel than other DNA fragments of identical length. The extent to which the bent DNA is retard ed is related to the pore size of the gel. A flexible DNA molecule that has no extensive stabilized bending will, by thermal motions, accommodate itself to the average pore size. By contrast, for a DNA molecule where stabilized bending increases the average radial dimensions, there will be a barrier to its progress through the gel; the DNA behaves similarly to a globular protein. Conversely, any stabilized bending that decreases the average radial dimension will in principle lower the resistance of the molecule to its passage through the pores and consequently will increase its rate of movement. The mobility of a stably bent DNA fragment depends not only on the direction of bending but also on the position of a bend within the fragment, a stable bend in the centre of the molecule having a much greater effect on mobility than a bend of the same magnitude close to the end of the molecule.

In kinetoplast DNA bending is dependent on a sequence which contains short runs of (dA).(dT) whose centres are spaced at intervals of

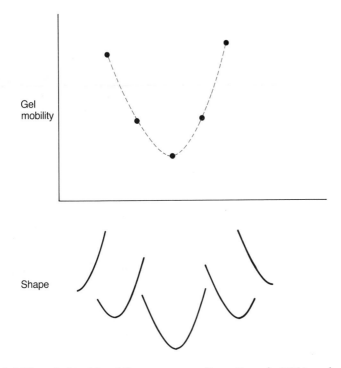

Figure 1.6 The relationship of the average configuration of a DNA molecule to gel mobility. The figure shows the effect of permuting a single DNA sequence so that the bend occupys different positions in the molecules. Note that the slowest moving species have the shortest distance between the two ends of the molecule.

10–11 bp, that is every double helical turn. Both the integrity of these short (dA).(dT) runs and their correct phasing are necessary for stable bending to be detected. Thus, sequences in which these homopolymer runs are regularly repeated every 8 or 12 bp do not exhibit anomalous mobility on polyacrylamide gel electrophoresis, whereas those with such a run repeated every ten base pairs do. This requirement for phasing can be understood in terms of the resultant direction of curvature of such sequences. When the sequence repeat is in phase with the double helical repeat (that is, they have the same periodicity), then any bending conferred by a single (dA).(dT) tract will be directed towards the same centre of curvature as that conferred by its neighbours (Figure 1.7) and consequently the DNA molecule will bend in a uniform curve. By contrast, when the helical periodicity is substantially different from the sequence periodicity, the direction of curvature conferred by one tract will differ from that of its neighbours and the cumulative curvature will be small or non-existent. A further significant property of (dA).(dT) tracts is the length of the tract required to confer intrinsic curvature. At normal temperatures a critical length of 4–5 bp is necessary; runs of 2 and 3 bp are insufficient to confer significant detectable curvature. More-over, if a short dA run is interrupted by any other naturally occurring base (C, G or T) intrinsic curvature is lost; only the replacement of an A-T base pair by the unnatural I-C base pair retains curvature.

The crucial structural basis for the conformational rigidity of (dA).(dT) tracts is the high propeller twist of the base pairs. In G-C base pairs rotation of the bases relative to each other is impeded by the presence of three Watson–Crick hydrogen bonds whereas in I-C base pairs, which have similar pattern of interbase hydrogen bonds to A-T base pairs, there is no such barrier. This property therefore explains why I-C base pairs can substitute effectively for A-T base pairs in (dA).(dT) tracts whereas G-C base pairs cannot. An additional consequence of the high propeller twisting is a narrow minor groove; in crystal structures the minor groove is unusually narrow with a minimum width of about 9 Å in contrast to an average width of 12–13 Å.

What property of these short (dA).(dT) tracts confers stable curvature on the DNA molecule as a whole? From the physical properties of intrinsically bent DNA molecules such tracts must be both relatively conformationally rigid and change the direction of the helical axis. Here it should be noted that the property of conformational rigidity does not imply stasis; rather the range of conformations easily adopted by the sequence is narrow. It has long been apparent that poly(dA).(dT) is structurally distinct from random sequence DNA, adopting a helical periodicity in solution of 10 bp per turn in contrast to the average of 10.5–10.6 bp per turn. Furthermore, uniquely, this homopolymer cannot be reconstituted into a nucleosome core particle and it is also exceptional in its failure to undergo the B to A transition in fibres. These two latter properties are indicative of conformational rigidity. A structural explana-tion for these properties was provided by the determination of the crys-tal structure of a DNA dodecamer containing a homopolymeric run of six A-T base pairs. In this crystal the (dA).(dT) tract is essentially straight, that is, the average planes through the base pairs are parallel to each other and perpendicular to the helical axis. Put another way, they have zero roll. In addition, the structure of the base pairs themselves is unusual. The propeller twist is high, about 20–25°. This results in max-imal overlap of the bases on each strand with a consequent increase in stacking energy and also the formation of a run of additional, non-Watson–Crick, cross strand hydrogen bonds. These bonds are located in the major groove and bridge the N6 position of adenine with the O4 of thymine. (Figure 1.7). These positions thus have the potential for forming hydrogen bonds in two directions; the bonds are bifurcated. Both the increased stacking energy and the bifurcated hydrogen bonds would be expected to confer a conformational rigidity over and above that

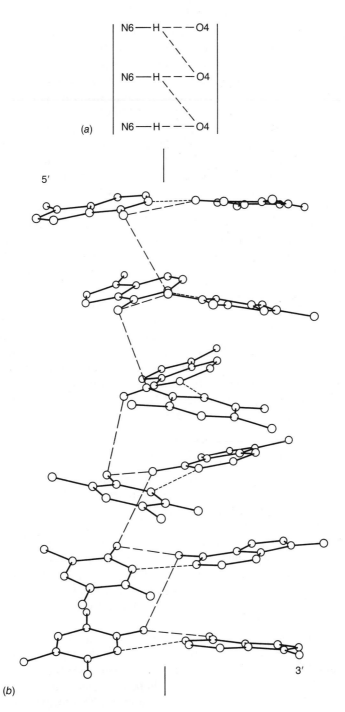

Figure 1.7 The structure of an oligo dA:dT tract. Bifurcated hydrogen bonding between N6 of adenine and O4 of thymine (a) is made possible by the propeller twisting of the A-T base pairs (b).

expected from base pairs with two hydrogen bonds. This structure also explains the length dependence of intrinsic bending. In any (dA).(dT) run only internal base pairs can participate in both enhanced stacking interactions and bifurcated hydrogen bonds with both neighbouring base pairs. The minimum length for a stable structure is thus 4 bp. The conformational rigidity of the oligo(dA).(dT) tract is also manifest in solution as a ten-fold lower rate of base pair opening within the tract.

The structure of (dA).(dT) tracts presents a paradox: although in crystal structures they themselves are straight, they confer curvature on a DNA molecule in solution. The property of curvature must thus arise from the fact that such tracts are structurally distinct from DNA of random sequence. In the crystal of d(CGCA$_6$GCG) the most abrupt changes in the direction of the local helical axis occur at the CA step and the GC step 3' to the dA tract. At both these steps the normal purine clash is exaggerated by the difference in propeller twist, resulting in a large positive roll angle for the pyrimidine–purine step and a corresponding negative roll angle for the purine–pyrimidine step. Much of the contribution to overall bending could thus arise from the large roll angles at the junctions of the A-T tract with flanking base pairs. Since the angles at the 5' and 3' ends of the tract are opposite in sign they would be additive when placed half a helical turn apart (Figure 1.8). However, it is not necessary to have all the change in direction concentrated at the junctions; it could be more widely distributed over the region between oligo(dA).(dT) tracts. To give a net bend there need only be a net roll accumulated in the intervening regions.

1.5 DNA SUPERCOILING AND TOPOLOGY

Inside the cell a long DNA molecule is not free to rotate about its long axis as a linear DNA molecule in solution would be. Instead, rotation is constrained either by proteins holding together a loop of DNA, as observed on the scaffold of eukaryotic chromosomes, or by the closure of a DNA molecule into a circle. Once rotation is constrained in this manner any physical changes which favour an alteration of the helical twist of DNA will place the molecule under torsional strain within the limits of the region in which free rotation is restricted. In closed domains, be they loops or circles, the DNA retains both axial and helical flexibility and, in general, any torsional stress will result in two forms of twisting: a change in the helical repeat, that is a change in the number of double helical turns, and also a twisting of the long axis of the DNA molecule. This axial twisting introduces a second type of turn, a writhe. Although they are physically distinct, the double helical 'turns' and the writhe are topologically equivalent in a closed domain. This means that, in the absence of any DNA strand breakage within a domain that would permit free rotation, the total number of writhing 'turns' and double helical turns must remain constant, and consequently, a change in one type of turn will be compensated for by a corresponding change in the number of the other type.

The principle of interconvertibility of writhing and double helical turns can be readily visualized with a short length of moderately flexible rubber tubing (Figure 1.9). Draw a line along the surface of the tubing parallel to the long axis. Then apply a twist at one end while holding the other end fixed. When stretched out the line will follow a helical path around the tubing; that is, the helical twist of the tubing has been changed from zero to some finite value. However, when the ends of the tubing are moved towards each other the axis of the tubing follows a helical path which has the same helical sense as the helical twist of the straight tubing. In this case the torsional stress is reflected principally in writhing turns resulting from the axial flexibility of the tubing. If the two ends of the tubing are now joined together the helical path of the long axis changes such that the torsional stress is now expressed as a twisting

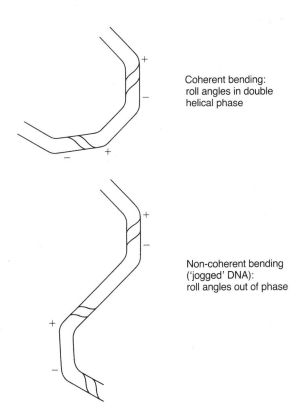

Coherent bending:
roll angles in double
helical phase

Non-coherent bending
('jogged' DNA):
roll angles out of phase

Figure 1.8 Coherent and non-coherent bending of DNA. Coherent bending results in macroscopic curvature; non-coherent bending produces a 'jogged' molecule.

of the long axis of the tubing about itself. These two modes of expression of the writhing turns are termed toroidal and plectonemic (or inter-wound) coiling, respectively. In a closed circular DNA molecule the plectonemic form is a lower energy state than the toroidal form. This is because in the former state any changes in the helical twist which are induced by the torsional strain are largely absorbed by the interwinding of the DNA double helices. Geometrically, if the DNA is negatively supercoiled, with the applied torsion being left-handed, then the res-ulting crossings in the plectonemic form will be right-handed (Figure 1.10). By contrast, in the toroidal form no such compensation can take place, and consequently it is the plectonemic form that is characterist-ically assumed by closed circular DNA molecules under torsional strain, whereas in many protein–DNA complexes the DNA is wrapped in a toroidal mode. The interwound form of superhelical DNA possesses one further property: unlike the toroidal form, which in principle can form a continuous helix of double stranded DNA, the interwound form must possess regions in which the DNA molecule bends in a loop to permit the winding of one part of the DNA duplex about the other (Figure 1.9). These regions, termed apices, are structurally distinct from the remain-der of the supercoiled molecule. If the physical properties of a closed circular DNA molecule were uniform along the double helix there would, in principle, be an infinite number of possible positions for the apices. However, it is likely that particular classes of sequence, for example intrinsically curved DNA, would prefer to be located at the apices rather than in the interwound duplexes. Such a preference would have two consequences: the structure of such a supercoiled molecule would be defined (and thus entropically disfavoured) and the config-uration of the DNA at the apices could serve as a recognition signal for DNA-binding proteins. The unique character of an apex also means that within such a region the partition between twist and writhe will be different from the remainder of the molecule and also the DNA will be in a higher energy state.

For a closed DNA domain we can define a third topological quantity, the linking number, Lk. The linking number is strictly defined as the number of times one chain crosses the other, a right-hand crossing being taken as positive. Defined in this way the absolute linking number of a domain is quantized and must be integral. In a supercoiled DNA mo-lecule the crossing of the DNA chains can occur in two ways, within the double helix itself and the crossings of the double helix on itself in the plectonemic form.

Since the toroidal and plectonemic forms of supercoiled DNA are interconvertible, their contributions to this latter term are equivalent. Expressed formally, the relationship between linking number, twist and writhe can be written:

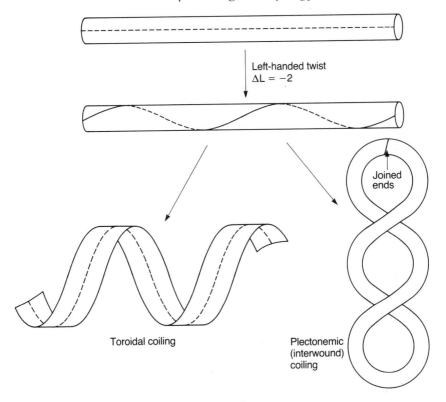

Figure 1.9 Toroidal and plectonemic coiling of DNA. In the manipulations the ends of the tubing are not free to rotate.

$$Lk = Tw + Wr$$

This equation represents the fundamental relationship in the topology of DNA. An alternative way of viewing the topological state of a DNA domain is to relate it to the relaxed state. In a relaxed molecule $Wr = 0$ and $Lk = Tw$, and consequently the change in linking number of the domain from the relaxed state may be expressed as:

$$\Delta Lk = \Delta Tw + \Delta Wr$$

The principal difficulty in the application of this equation is the use of the correct reference frame for the definition of twist and writhe. The twist of a DNA molecule can be defined in two ways. The absolute or intrinsic twist (also termed laboratory twist) is a measure of the average angle between successive base pairs and is thus related to the local conformation of the base steps. It is this quantity which is commonly understood as the helical repeat. By contrast, the relative (or local twist)

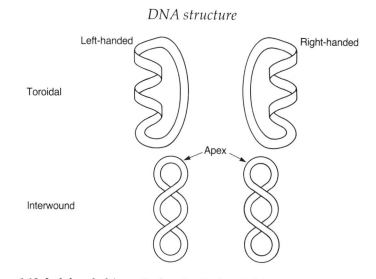

Figure 1.10 Left-handed (negative) and right-handed (positive) configurations of supercoiled DNA.

is defined with respect to the surface on which a DNA molecule is wrapped and is a measure of the number of base pairs between geometrically equivalent positions on the DNA double helix. This concept may be simply understood by considering a DNA molecule wrapped as a uniform superhelix around a cylindrical surface. For such a molecule the local twist is the number of base pairs between, for example, successive outward facing minor (or major) grooves. In such a superhelix the local twist does not correspond to the intrinsic twist, as in the former reference frame there is an additional twist of $2\pi \sin \alpha$ associated with one turn of the superhelical path, where α is the pitch of the superhelix. This is distributed equally over the number of bases (N_b) in one turn of the superhelical path so that:

$$Tw_i = Tw_r + (2\pi \sin \alpha)/N_b$$

From this relationship it follows that, if the intrinsic twist remains constant, then the relative twist will determine the three-dimensional path of the double helical axis. When these two quantities are identical $(2\pi \sin \alpha)/N_b$ will be zero and the DNA molecule will adopt a circular form (Figure 1.11). By contrast, when $Tw_r > Tw_i$ the DNA will assume the form of a left-handed superhelix, and in the converse situation a right-handed superhelix. The writhe of a DNA molecule is defined in the same reference frame as the absolute twist. It is not the same quantity as the number of superhelical turns in a uniform superhelix. This latter quantity is defined in the local frame and again can be related by simple geometry to the true writhe.

The relationship between the intrinsic and relative (or local) twist has one further consequence which is relevant to the ability of proteins or protein complexes to alter the intrinsic twist of DNA, that is to unwind or overwind the double helix. This is best illustrated by the Indian rope trick. Imagine a DNA molecule (or a rope) wrapped as a shallow left-handed superhelix in the form of a cylinder (Figure 1.12). One end of this superhelix is then pulled out so that the number of superhelical turns remains constant but the pitch, α, of the superhelix alters. When this operation is performed the length of a superhelical turn remains constant as does the number of base pairs it contains. The relative twist, which in this situation is independent of α, must also necessarily remain constant. However, the intrinsic twist and the writhe are both dependent on $\sin\alpha$. Therefore, as the supercoil is pulled out so the intrinsic twist decreases (because in a left-handed superhelix α is negative) and the DNA becomes structurally underwound. At the same time, the left-handed writhe decreases (i.e. the writhe becomes less negative). The structural consequence of pulling out the supercoil is thus a direct interconversion of twist and writhe. This principle can be applied to any DNA molecule in a toroidal form. Note that the important conversion is a change in the pitch angle of the toroid; it is not a necessary condition that the super-helix be pulled out and maintain a cylindrical form. An equivalent conversion could be achieved by a change in shape of the toroidal from, for example, a capstan to a barrel (Figure 1.13). In such a situation the change in the geometry of the bound DNA means that the normal of the wrapped DNA to the surface is no longer perpendicular to the overall

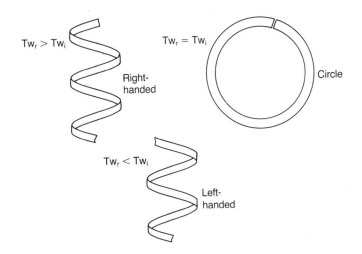

Figure 1.11 The relation between the intrinsic and relative twist determines the configuration of a DNA molecule.

superhelical axis. Alterations in intrinsic twist induced by changes in DNA configuration of the type discussed here will in principle be distributed relatively uniformly along the DNA molecule, subject only to local variations in the ability of DNA to unwind. The precisely localized unwinding mediated by proteins requires additional constraints.

1.6 THE TOPOLOGY OF PROTEIN-BOUND DNA

When a DNA molecule is bound to a protein it may differ in its topological state from free DNA. In effect, the complex can be regarded as a small closed topological domain. Only when the protein is removed from the DNA does this short domain become part of the larger domain of the whole DNA molecule. This release of the topological properties of DNA can be made use of experimentally to determine the extent of topological unwinding or overwinding in a protein–DNA complex. The principle of the experiment is simple, and is illustrated in Figure 1.14. The protein is first bound to a specific site on a long DNA molecule. The two ends of the DNA are ligated together to form a covalently closed circle containing, on average, no superhelical turns. Once the ring is closed the protein can be removed. Any transfer of topological under- or overwinding in the protein–DNA complex will change the superhelicity of the DNA circle. The extent of change of superhelicity is then measured in polyacrylamide gels, in which the rate of movement of circular DNA molecules depends on the number of superhelical turns they contain. In this way the extent of topological winding in the protein–DNA complex can be accurately determined.

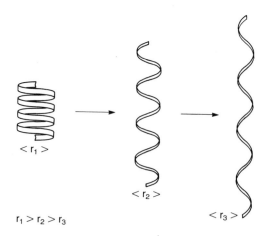

$r_1 > r_2 > r_3$

Figure 1.12 The Indian rope trick.

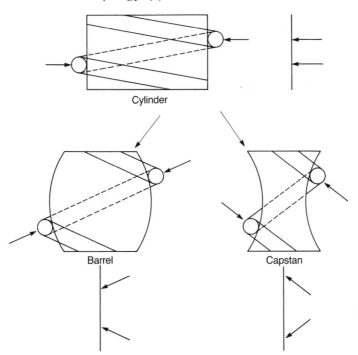

Cylinder

Barrel   Capstan

Figure 1.13 The relation of the direction of curvature of a DNA superhelix to the superhelical axis in a cylinder, barrel and capstan.

A good example of the principles involved is provided by the nucleosome core particle. From the structure of crystals of this particle we know that 145 bp of DNA are wrapped around a histone octamer in about 1.75 left-handed superhelical turns. Yet, when the protein is removed, only one turn, and not 1.75, is recovered in the circular DNA molecule. This means that the topological unwinding of DNA in the nucleosome core particle is approximately one left-handed turn. The remaining expected 0.75 left-handed turn must therefore be cancelled by some other change in the topology of the protein–DNA complex. The answer to this apparent paradox is that the DNA on the nucleosome is overwound, that is, it has a local twist which differs from the value which it would normally assume in solution. For a DNA molecule wrapped in a uniform curve, the local twist is identical to a sequence periodicity which reflects the regular periodic changes in the local conformation of wrapped DNA. This periodicity, and hence the local helical repeat, for nucleosome core DNA is 10.2 bp. By contrast, in solution, the average helical repeat of this same DNA is 10.6 bp. In other words, on the nucleosome 145 bp of bound DNA

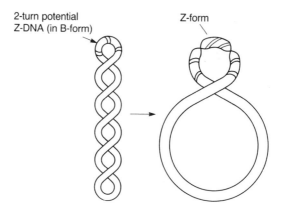

2-turn potential
Z-DNA (in B-form)

Z-form

Figure 1.14 Topological consequences of a B → Z-DNA transition.

will contain 145/10.2 = 14.2 right-handed turns, whereas in solution, free of protein, this same DNA will contain 145/10.6 = 13.7 double-helical turns. The difference of (14.2 − 13.7) = 0.5 turns accounts in large part for the paradox. The total change in linking number on nucleosome formation can therefore be written as:

$$\Delta L = -1.75 \qquad\qquad +0.5 \qquad\qquad\qquad = -1.25$$

Number of	Excess of	Net topological
left-handed	right-handed	change
superhelical turns	double helical turns	

Both as measured in the local reference frame

A further consequence of the wrapping of DNA in a superhelix around a protein core is that the formation of such complexes will be energetically favoured when assembled on DNA molecules which are themselves supercoiled in the same sense. For example, the formation of nucleosomes, of RNA polymerase–promoter complexes and of protein complexes involved in site specific recombination, all of which contain DNA wrapped in a left-handed sense, is facilitated by negatively super-coiled (i.e. underwound) DNA.

1.7 STRUCTURE OF SUPERCOILED DNA

Many enzymatic manipulations of DNA molecules require changes in the helical twist. Thus the crucial processes of DNA replication, DNA transcription and site-specific recombination all necessitate a local un-winding of the DNA double helix. It follows that any torsional strain which favours unwinding will also facilitate these processes. In both eubacterial and eukaryotic cells, DNA which is available to the appropriate enzymes is under such torsional strain, which is normally in the opposite or negative sense to the helical twist of DNA. DNA molecules in this state are at a higher energy level than torsionally unstrained, or relaxed, molecules. The average structure of the double helix is consequently different in the supercoiled and relaxed states. In particular, the supercoiled state permits the formation of DNA structures which would be unstable in relaxed molecules.

A convenient measure of supercoiling is the superhelical density, θ, defined as Lk/Lk_0. Typically closed circular DNA molecules isolated from bacterial cells have a superhelical density in the region of −0.06, that is, one superhelical turn for approximately every 16–17 double helical turns. There are two important considerations with regard to this number. First, it is an average value and therefore need not reflect the actual superhelical density in any given region of the DNA molecule *in vivo*. Secondly, in the bacterial cell the measured superhelical density of the isolated free DNA does not represent the superhelical density available *in vivo* for promoting conformational and topological transitions in DNA structure. This is because in the bacterial chromosome a certain fraction of the total superhelical turns observed in the isolated DNA will be constrained by wrapping on a protein surface. Indeed, careful estimates of the available, or unconstrained, superhelical density suggest that at least half the superhelical turns are locked in this way. In addition, even the unconstrained observed superhelical density of a given isolated DNA molecule is an average value; locally there may be substantial variation in the superhelical density away from the average, generated by such processes as transcription or by helix-tracking proteins.

This variation in superhelical density means that locally this density may be sufficiently high to allow the DNA molecule to assume structures that differ from the normal right-handed duplex and which require a high activation energy for their formation. The formation of such structures is almost invariably driven by high negative superhelical densities and results in a reduction in the average superhelical density in the remainder of the domain. The three principal structures of this type so far described are Z-DNA, cruciforms and H-DNA.

Z-DNA is a double helical form which is left-handed and contains 8 bp per double helical turn. The formation of this structure is strongly dependent on the DNA sequence. In particular sequences of the type (dCG).(dCG), or any alternating purine–pyrimidine sequence, possess the lowest energy barrier to overcome the B to Z transition. The importance of such a structure in relieving the torsional strain over the bulk of a topological domain can be appreciated from the consideration that the formation of a single turn of left-handed Z-DNA will reduce this strain in the remainder of the domain by +2 (right-handed) turns (Figure 1.14). The existence of two double helical stems in a cruciform structure is not obligatory. Certain DNA sequences from eukaryotic telomeres which have the form (GGGGAAAT)$_n$ will, at low pH, form a cruciform-like structure under negative superhelical strain. In this structure there is only one base-paired stem in which the pairing is not of the normal Watson–Crick type, while the complementary strand has no apparent secondary structure (Figure 1.15).

The formation of cruciforms and H-DNA again requires negative superhelical strain. The classic cruciform structure requires a palindromic sequence which allows the formation of a local base-paired region with a single DNA strand as an alternative to the normal duplex. H-DNA, by contrast, involves a different mode of base-pairing. This structure is formed by sequences containing runs of (dCT).(dAG) or (dG).(dC) and is characterized by the acquisition of asymmetric

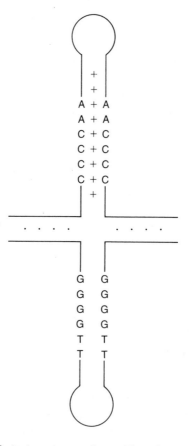

Figure 1.15 The (C.A) hairpin structure formed by telomeric DNA under superhelical strain.

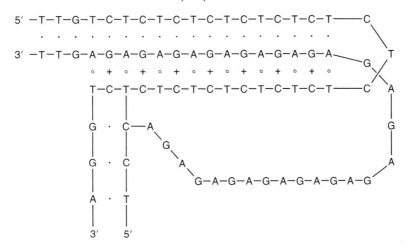

Figure 1.16 The triple stranded H-form of DNA. The triplex includes a normal Watson–Crick duplex (●) associated with a pyrimidine loop through Hoogsteen base pairs (o, +).

sensitivity to cleavage reagents which sense single DNA strands. This observation suggests that within H-DNA one DNA strand, but not its complement, is freely accessible to such reagents. The proposed structure for HDNA is thus a region of triple stranded DNA in which a strand of the repeating sequence is wrapped in the major groove of a duplex of the same sequence (Figure 1.16). In this configuration the third strand can form Hoogsteen base pairs[*] with the strands in the duplex that are parallel to itself, thereby stabilizing the structure (Figure 1.17). The existence of such a triple strand automatically means that the complement to the wrapped strand will not be involved in base pairing. Triple stranded H-DNA is essentially an asymmetrical structure which implies that, provided the Hoogsteen base-pairing requirements can be satisfied, two isomers should, in principle, exist. For the homopolymer (dG).(dC) this is indeed the case, the differences in the stability of the two forms being dependent on the pH and the presence of divalent cations. The pH effect can be directly related to the protonation of cytosine residues which allow the formation of Hoogsteen base pairs in one isomer but not in the other.

Yet another variant of DNA structure is a four stranded quadruplex. Structures of this type are typically formed by telomeric DNA which

---

[*]Hoogsteen base pairs utilize different hydrogen bonds from Watson–Crick base pairs. For an A–T base pair to the Watson–Crick pair hydrogen bonds are formed between A–6NH$_2$ T–4O and A–N5 to T–3NH; for the Hoogsteen pair the bonds are formed between A–6NH$_2$ to T–4O and A–N7 to T–3NH.

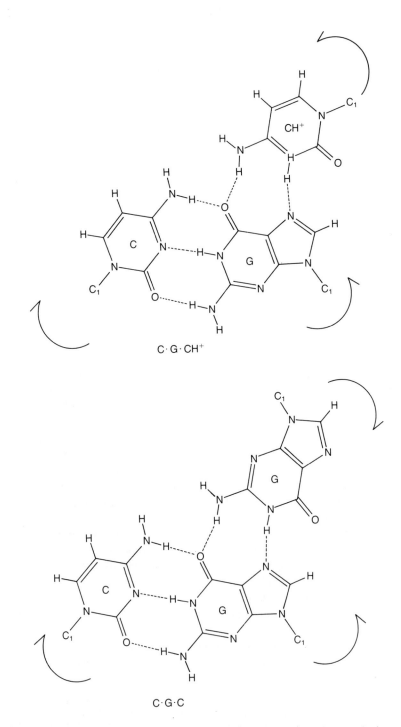

Figure 1.17 Base triplets formed in H-DNA. The C.G.CH$^+$ triplet is only formed at low pH.

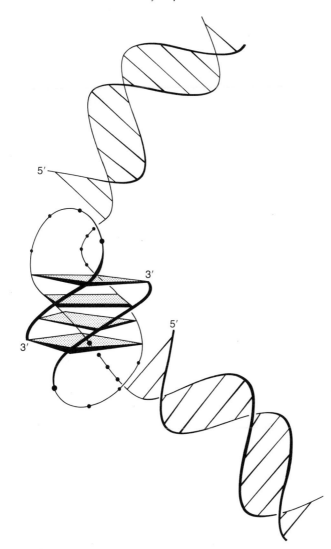

Figure 1.18 (a) The structure of a quadruple helical DNA formed by telomeric DNA.

occurs at the ends of linear eukaryotic chromosomes. Most, but not all, telomeres consist of repeated tracts of ~3–8 guanine residues separated by short tracts of A or T. When single stranded the guanine residues can form cyclic base tetrads, which can then stack on each other such that the strands in the tetrad region are arranged in an antiparallel fashion. The A-T rich linking sequence then forms a loop linking the strands (Figure 1.18). Again the biological role, if any, of such structures remains to be established.

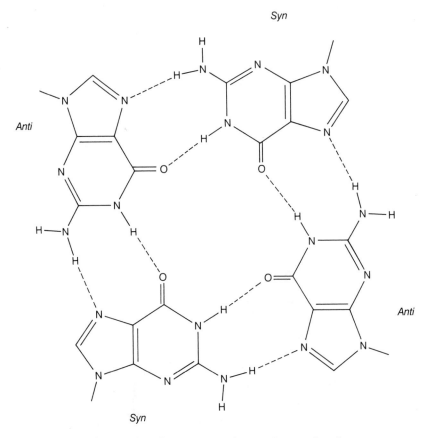

Figure 1.18 (b) The structure of a guanine quadruplex.

REFERENCES

Dickerson, R.E. and Drew, H.R. (1981) Structure of a B-DNA dodecamer. *Journal of Molecular Biology*, **149**, 761–86.

Laundon, C.H. and Griffith, J.D. Curved helix segments can uniquely orient the topology of supertwisted DNA. *Cell*, **52**, 545–9.

Nelson, H.C.M. *et al.* (1987) The structure of an oligo(dA)-oligo(dT) tract and its biological implications. *Nature*, **330**, 221–6.

Sundquist, W.I. and Klug, A. (1989) Telomeric DNA dimerizes by formation of quanine tetrads between hairpin loops. *Nature*, **342**, 825–9.

Wang, J.C. *et al.* (1979) Molecular structure of a left-handed double helical DNA fragment. *Nature*, **282**, 680–6.

Voloshin, V. *et al.* (1988) Palindromic structure of H-DNA. *Nature*, **333**, 475–6.

Williamson, J.R., Raghurarman, M.K. and Cech, T.R. (1989) Monovalent cation-induced structure of telomeric DNA: the G-quartet model. *Cell*, **59**, 871–80.

Wu, H-M. and Crothers, D.M. (1984) The locus of sequence-directed and protein-induced DNA bending. *Nature*, **308**, 509–13.

2

DNA–protein interactions: The three-dimensional architecture of DNA–protein complexes

2.1 GENERAL PRINCIPLES

Before a gene can be transcribed into RNA or site specific recombination can take place a specific binding site on the DNA must be recognized by a protein. The specific positioning of a protein or protein complex on a particular DNA sequence requires that there be more favourable interactions between the protein and its binding site than between the same protein and an unfavourable DNA sequence. The selective binding of a protein to a particular DNA sequence requires the recognition by the protein of an ensemble of steric and chemical features that in total delineate the binding site. This process of selective recognition is essentially comparable to the binding of any ligand, be it enzymatic substrate or allosteric effector, to a protein molecule. In the particular case of protein–DNA interactions, the original models for selective recognition invoked principally hydrogen bonding interactions between the proteins and the individual bases, the remainder of the DNA molecule being considered to lack sufficient information for selectivity. However, it has become apparent that both the local conformation and the local configuration of a DNA molecule can have a profound influence on protein–DNA interactions. In some instances, notably for the histone octamer and for the *E. coli trp* repressor, these factors can by themselves act as the major, and possibly the sole, determinants of selectivity. This type of recognition has been termed indirect read-out or analogue recognition in contradistinction to the direct or digital recognition of individual bases or base pairs. In general however, both modes of recognition contribute to the specificity of the complex and the available examples constitute a spectrum of the relative contributions of these components.

The trajectory of the DNA in its complexes with protein plays an important role in the biological function of DNA. Such bending can be intrinsic to the DNA, or can be induced by bound protein, either as a free loop between proteins bound at separated sites or, alternatively, spooled on the protein surface. Additionally it should be recalled that bending, as writhe, is a concomitant of supercoiling, either positive or negative. In all these examples the bending of the DNA requires a pattern of local conformations that differ from those for 'straight' DNA.

2.2 LOCAL DNA CONFORMATION AND PROTEIN BINDING

Many proteins, of which nucleases are the best example, interact with a relatively short region of DNA with low selectivity. In two such cases there is now substantial evidence that a particular local conformation of the DNA is necessary for a productive interaction.

One of the best studied examples of this type of interaction involves bovine pancreatic deoxyribonuclease I (DNase I). This enzyme, under standard conditions, introduces a single scission in one strand of the DNA double helix. However, the cleavage rates of this enzyme vary along a given DNA sequence, indicating that the enzyme is sensitive to structural variations in the double helix. In particular, DNase I cleaves DNA at a low rate at homopolymeric tracts of (dA).(dT) and (dG).(dC) relative to 'random' sequence DNA. Under normal digestion conditions the enzyme cleaves the DNA sugar–phosphate backbone so that the separation of the independent cleavage sites on the complementary strands is 2–4 bp, a value consistent with recognition of the minor, but not the major, groove of DNA. Two other properties of DNase I are relevant to its cleavage selectivity. First, when presented with a small closed circular DNA with a high radius of curvature (< 150 Å) the enzyme preferentially cleaves positions on the DNA where the minor groove is on the outside of the circle but not positions where the minor groove is on the inside. Secondly, when a DNA molecule is tightly wrapped on a protein surface an exposed minor groove is often cleaved at a much higher rate than the same sequence in free DNA.

The importance of the minor groove in the mechanism of action of DNAase I is apparent from the crystal structure of the enzyme bound to an octanucleotide. In this structure an exposed loop of the enzyme binds to the minor groove of DNA, such that contacts are made with both flanking sugar–phosphate backbones while a tyrosine and an arginine residue penetrate into the groove itself (Figure 2.1). The most striking feature of the complex is that the DNA is deformed, showing a 21.5° bend towards the major groove and away from the bound enzyme. This bend results in a widening of the minor groove to 15 Å from an average width of 11–12 Å. The bend is also localized, the principal change in the

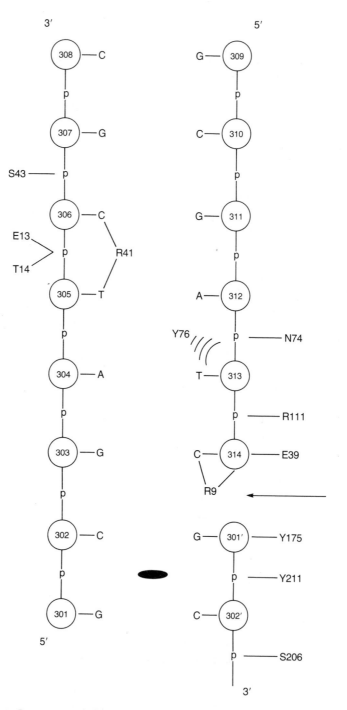

Figure 2.1 Contacts made by DNase I to bases and to the sugar–phosphate backbone in a crystal of the enzyme with a DNA oligomer of defined sequence, as viewed in a projection of the minor groove of DNA.

direction of the helical axis occurring at the base step immediately adjacent to that where the cleavage is located (Figure 2.2).

The major inference from this structure is that, in order to cleave a DNA molecule, DNase I must first bend it to the appropriate conformation. It follows that any DNA sequence that cannot be readily deformed to the required conformation will show a reduced rate of cleavage while, conversely, any DNA segment already stabilized in the preferred conformation will exhibit an enhanced susceptibility. In the former category are sequences such as oligo(dA).(dT), and to a lesser extent oligo(dG).(dC), which by virtue of their base stacking interactions are believed to be conformationally rigid. Similarly, the minor groove on the inside of small DNA circles would be less sensitive to DNase I because the local curvature of the DNA is constrained to the opposite direction to that required for cleavage (Figure 2.3). By contrast, when DNA is wrapped on the surface of a protein, as in the nucleosome core particle,

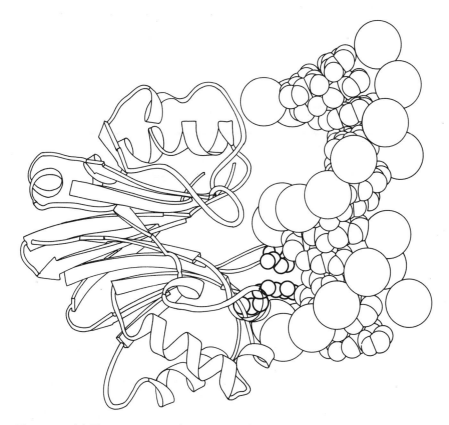

Figure 2.2 (a) The interaction of DNase I with DNA. Note that the major contacts are across and within the minor groove.

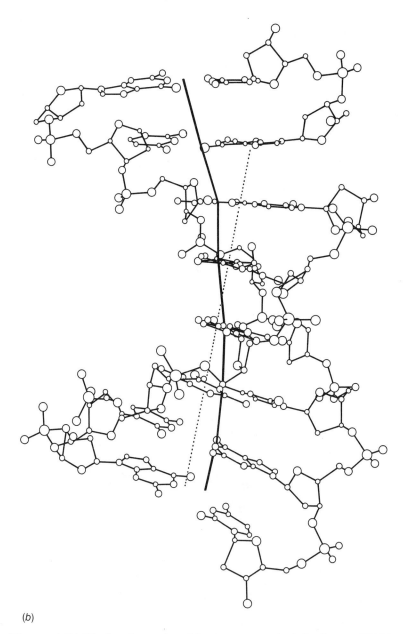

(b)

Figure 2.2 (b) The bend introduced by DNase I in the helical axis of DNA.

the direction of curvature is such that an outward-facing minor groove could be in the correct orientation for rapid cleavage. For DNase I, comparison of the crystal structures with and without bound DNA shows that the protein itself is not significantly deformed by interaction with

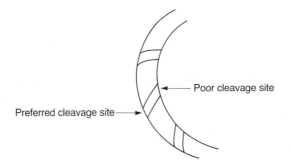

Poor cleavage site

Preferred cleavage site

Figure 2.3 Selectivity of DNase I cleavage. The enzyme cleaves the minor groove where the axis of the DNA bends away from the protein.

its substrate. Thus the probability of cleavage at a particular base step is likely to be a measure of the ability of that base step and its immediate neighbours to assume the appropriate conformation. In other words, the selectivity of cleavage by DNase I is dependent in part on the flexibility of the DNA in the immediate vicinity of the cleavage site.

A second example of an enzyme that is dependent on the local conformation of DNA for function is topoisomerase I. In eukaryotes this protein relaxes both positive and negative supercoils in contrast to the prokaryotic enzyme which only relaxes negatively supercoiled DNA. For both enzymes the relaxation proceeds by the introduction of a break in one strand of the sugar–phosphate backbone, thus allowing the free rotation of one region of the double helix relative to the other. In the case of the eukaryotic enzyme both this cleavage and the subsequent relaxation are dependent on the superhelical density. However, the initial binding of the topoisomerase to the DNA is relatively insensitive to this parameter. These properties show that the enzyme must sense the torsional strain in the DNA, presumably by the recognition of a strain-induced conformational feature. Torsional strain can affect conformation by changing both the twist of the DNA and the average bending ('writhe'). Whereas the nature of the changes in twist are strongly dependent on the sense of the torsional strain, those induced by writhing will depend more on the local geometry, that is on the direction of bending. That this latter property is crucial for enzymatic function is shown by that observation that the eukaryotic enzyme will efficiently utilize intrinsically bent, yet relaxed, DNA as a substrate. With this DNA the cleavage sites occur adjacent to the oligo(dA).(dT) tracts which confer the intrinsic bending. For prokaryotic topoisomerase I cleavage occurs principally in the vicinity of the easily unwound TpA steps, a property which is consistent with the notion that this enzyme senses torsional strain by identifying the transient base pair opening at such sites.

2.3 DNA WRAPPING

In many DNA–protein complexes the DNA is tightly wrapped around a central core of protein. The archetypal example of such a complex is the nucleosome core particle in which 145 bp of DNA are wrapped around the histone octamer in a superhelix of 43 Å radius (Figure 2.4). This particle is the basic structural unit of eukaryotic chromatin and is associated *in vivo* with an immense variety of DNA sequences. Yet in the test tube the histone octamer can position itself precisely with respect to a defined DNA sequence. The ability of the protein complex both to position itself precisely and to associate in this way with a wide variety of sequences shows that sequence specific recognition is not a significant determinant of nucleosome positioning. Selectivity is thus likely to depend on a more diffuse or mechanical property of the DNA. The tight bending of the DNA in the nucleosome suggested that the ability of the DNA to bend in a particular direction (its bendability) might be a crucial factor. To test this hypothesis experimentally it must be established that the bending preference of a particular DNA molecule is the same when reconstituted into a nucleosome as it is when constrained in the absence of any bound proteins. To approach this question, a 169 bp DNA fragment of bacterial origin was closed into a small relaxed circle. Such a molecule, if bent uniformly, would have an average diameter of about 170 Å, that is, about twice that of the DNA superhelix on the nucleosome core. The preferred direction of curvature, or the rotational orientation of the DNA sequence, was established using DNase I as a probe, assuming that the enzyme would preferentially cleave the DNA where the minor groove was exposed on the outside of the circle. The resulting cleavage pattern demonstrated that the particular DNA sequence used adopted a highly preferred orientation, or, in other words, the molecule was bent in one direction and not in others. When this same DNA molecule, now in linear form, was reconstituted into a nucleosome core particle, the direction of bending of the DNA remained largely conserved in going from circle to nucleosome, so that where the minor groove was on the outside of the circle it was also on the outside of the superhelix on the nucleosome core particle (Figure 2.5). The few slight differences that were observed can be attributed to an alteration in the average cleavage periodicity (which reflects the average twist of the DNA molecule) from 10.56 bp in the circle to 10.32 bp in the nucleosome core particle. This result is fully consistent with the view that DNA bendability is a major determinant of nucleosome positioning.

When DNA is tightly bent the outer circumference of the molecule will be considerably larger than the inner circumference. In the case of the nucleosome core particle the external circumference of a single super-helical turn is 333 Å while the internal circumference is only 207 Å. This substaintial difference requires that, on the exterior surface of the

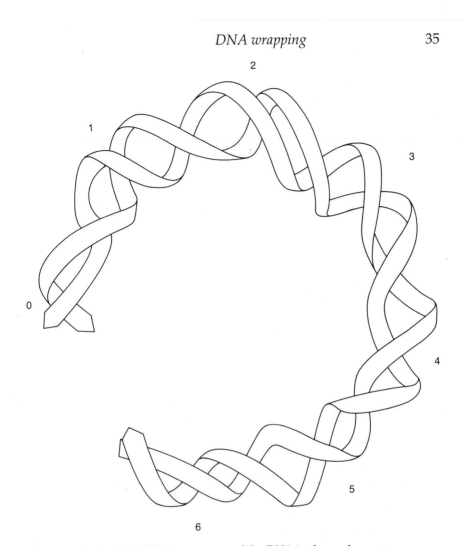

Figure 2.4 Path of the DNA in one gyre of the DNA in the nucleosome core particle.

superhelix, the grooves of the DNA, both major and minor, be expanded, while those on the inside are compressed (Figure 2.6). Thus, to follow such a superhelical path the bound DNA must be able to accommodate the deformations associated with bending. The ability of the DNA to assume the required conformations is sequence dependent, and thus the organization of the DNA sequence in a particular region will determine the bendability of that region.

If the rotational positioning of DNA on a nucleosome is influenced by particular sequences, it would be expected that, in a population of nucleosome core particles, the occurrence of such sequences should, on

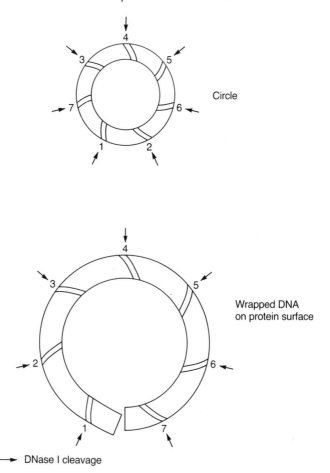

Circle

Wrapped DNA
on protein surface

→ DNase I cleavage

Figure 2.5 The orientation of a DNA molecule in a small circle is conserved when the molecule is wrapped around a histone octamer.

average, exhibit a periodic modulation that would reflect the structural periodicity of the DNA lying on the surface of the histone octamer. By the use of two techniques, statistical sequencing and direct sequencing, the nature of the short DNA sequences associated with the bending of nucleosomal DNA was established. In the method of statistical sequencing the predominant locations of particular DNA sequences in a mixed population of aligned DNA molecules of approximately equal length (145 bp) are probed by the binding of drug molecules of known sequence specificity to the DNA and then detecting the preferred binding locations by digestion of the drug–DNA complexes with DNase I (Figure 2.7). When applied to nucleosome core DNA from chicken

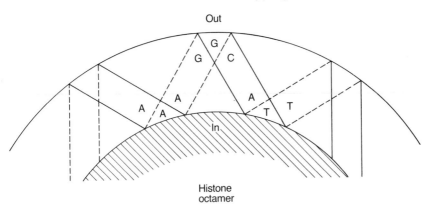

Figure 2.6 Preferred orientation of A/T and G/C rich sequences in DNA wrapped on a protein surface.

erythrocytes this technique showed that the occurrence of short runs of both d(A.T) and d(G.C) is periodically modulated with an average period of about 10.2 bp. Significantly, these two modulations are in opposite phases such that when d(A.T) runs are abundant d(G.C) runs are scarce and vice versa. We know from the crystal structure of the nucleosome core particle that at the mid-point of the DNA binding site (the 'dyad') the minor groove of the DNA points away from the protein core. Therefore, from a knowledge of the phase of the modulation of the d(A.T) and d(G.C) runs relative to the mid-point of the core DNA sequence it could be deduced that in the core particle the minor groove faces approximately inward towards the histone octamer, while d(G.C) runs prefer to occupy positions where the minor groove points outwards. This result again strongly supports the idea that in the nucleosome core particle the orientation of the DNA molecule relative to the protein surface is determined principally by certain directional bending preferences of the DNA, rather than by any sequence specific protein–DNA contacts.

Although statistical sequencing allows a description of the most dominant sequence features of nucleosome core DNA it cannot describe the occurrence of all particular sequence combinations; nor can this technique assess the detailed nature of the sequence preferences in regions such as those close to the dyad where the path of the DNA, instead of following a uniform superhelical path, 'jogs' to produce an S-bend linking two superhelical turns (Figure 2.8). To investigate these aspects of nucleosome core structure 177 individual core DNA molecules from the same DNA sample that had been analysed by statistical sequencing were cloned and sequenced by conventional methods. These sequences

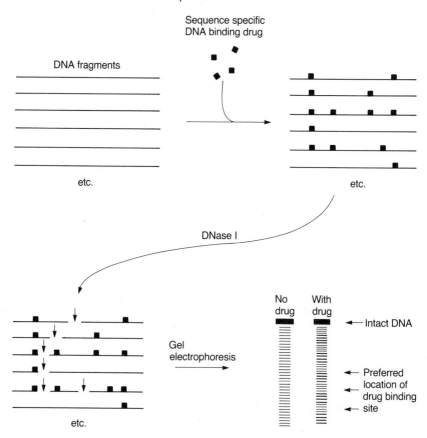

Figure 2.7 Statistical sequencing. The specificity of DNA-binding drugs is used to identify preferred positions of A/T and G/C rich sequences in a population of DNA molecules of approximately equal length.

were then aligned about their mid-point and analysed for the occurrence of dinucleotides and trinucleotides. The most striking result from this analysis is a well defined modulation in the occurrence of dinucleotide AA and its complement TT, which is interrupted only in the region of the dyad (Figure 2.9). The average spacing between positions of maximum abundance of these dinucleotides is 10.2 bp. A second dinucleotide, GC, also exhibits a strong periodic modulation but its phase, as deduced from Fourier analysis, is opposite to that observed for AA and TT. Thus GC must preferentially occur at positions on the DNA molecule which are separated by half a double helical turn from the preferred positions for AA and TT.

It is clear from the crystal structures of short DNA fragments that the conformation of a particular base step depends not only on its

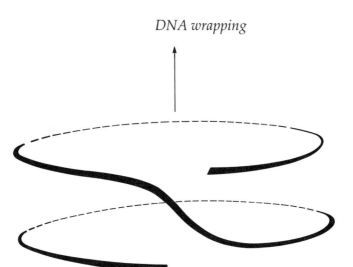

Figure 2.8 The path of the DNA in the nucleosome core particle. The arrow identifies the superhelical axis.

constituent base pairs but also on the sequence context in which it is located. This context effect is clearly apparent in nucleosome core DNA. The periodic modulation of the dinucleotides AA and TT reflect principally more pronounced modulations of the trinucleotides AAA/TTT and AAT/ATT, while other trinucleotides containing AA (or TT), such as CAA/TTG, do not occupy any strongly preferred positions with respect to the protein surface. Similarly, the modulation of the dinucleotide GC reflects largely the modulation of the trinucleotide GGC/GCC, and to a lesser extent that of AGC/GCT.

Several points should be emphasized with respect to these findings. First, the data clearly show that rotational positioning depends on sequence preferences and not absolute sequence requirements. In any piece of mixed sequence DNA it is unlikely that all the bending preferences in that sequence can be satisfied simultaneously by the configuration assumed by the whole molecule. Rather, the overall rotational setting will be determined by the balance of local preferences. Consequently, there will always be a few segments of helix for which rotational position is imposed not by local constraints but by the preferred configuration of the whole molecule. Secondly, the structural code for rotational positioning is redundant. In any bent DNA the conformational requirements at an inward- or outward-facing minor groove can be satisfied by several trinucleotides and their complements. This implies that, for any DNA sequence which must both encode a protein and have the potential to be tightly bent, the coding function will be the principal determinant of the sequence. Thirdly, rotational positioning by itself is insufficient, in prin-

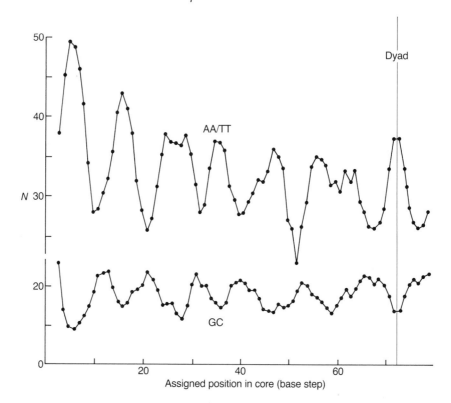

Figure 2.9 The distribution of the dinucleotides AA/TT and GC in core nucleosomal DNA. In the plot shown the number of occurrences has been symmetrized about the mid-point (dyad).

ciple, to specify a precise, or unique, location for a histone octamer on a defined DNA sequence. Such specification requires sequence markers that occur non-periodically and would be expected to be found either at unique locations within the nucleosome, such as the dyad, or at positions asymmetrically arranged with respect to the dyad. More detailed analysis of nucleosome DNA sequences reveals precisely these features. The sequence preferences at the dyad differ from those at any other location in nucleosomal DNA while the average sequence organization of the four double helical turns spanning the dyad is asymmetrically arranged with respect to the mid-point.

The sequence preferences for rotational positioning determined for nucleosomal DNA are fully consistent with the sequence dependence of DNA structure observed in crystals of short DNA oligomers. In such crystals A-T rich sequences (whether alternating or homopolymeric) but not G-C rich sequences, have a strong tendency to adopt a narrow minor groove. This observation suggests that the compression of the minor groove on the inside of a tightly bent DNA molecule can be accommodated more easily by A-T rich sequences than by G-C rich sequences. Conversely, the trinucleotide GGC in the crystal structure of d(GGGGCCCCC) has a large total positive roll of +20° that opens the minor groove, and thus such sequences occur preferentially at an outward facing minor groove.

A crucial question is to what extent the sequence preferences observed on the nucleosome also apply to other protein–DNA complexes containing bent DNA. In two cases, that of the bacteriophage 434 repressor–operator complex and that of the catabolite activator protein (CAP) interaction with the *lac* regulatory region, there is sufficient information available to answer this question. Both these complexes are examples of interactions in which the conformational flexibility of DNA complements identified, or assumed direct recognition of, the base sequence.

The complex of CAP with its binding site in the *lac* control region is of particular interest. The existence of significant DNA bending has been inferred from the anomalous mobility of the complex in polyacrylamide gel electrophoresis and from electron micrographs of the complex. In the complex the strong binding interactions span 28–30 bp of DNA as indicated by ethylation interference experiments. However, sequence conservation in the CAP binding sites is essentially confined to the central double helical turn. Within this region there is an interrupted inverted repeat with potential major groove contacts half a double helical turn on each side of the centre of symmetry. Thus, at the centre of the binding site the minor groove of DNA must point towards the protein. At this point and at 10 bp on either side are A-T rich sequences, which on the nucleosome occur preferentially where the minor groove is narrowed. Similarly, exactly out of phase are sequences that preferentially occupy an outward facing minor groove (Figure 2.10). Two conclusions can be drawn from this precise correspondence between the CAP binding site and nucleosomal DNA. First, the sequence of the binding site suggests that DNA bendability contributes significantly to the stability of the CAP–DNA complex. Secondly, the sequences involved in specific recognition at the major groove contacts, that is GTG and CAC, are also compatible with the inferred direction of bending in the complex.

Exhaustive mutagenesis of two particular regions within the binding site where the minor groove of DNA points respectively toward and away from the protein surface shows that mutations in these regions affect both the affinity of CAP for the binding site and also the magnitude of DNA bending in the complex, as indicated by its gel mobility. Strikingly, correlation of the extent of the inferred bending with the sequence changes parallels that deduced from nucleosomal DNA. This result indicates that the sequence dependence of protein induced DNA bending is a general property of the DNA itself and is not imposed by the binding protein.

Although the bendability of DNA affects both the affinity and selectivity of a protein for its binding site the energy required to deform DNA mildly in a non-preferred direction is small and is a good deal less than that available for direct sequence recognition (for example, by hydrogen bonding). Nucleosome reconstitution studies with defined DNA sequences suggest that, at each site where bending occurs (that is, where the minor groove points out or in), the difference in binding energy resulting from the substitution of a preferred for a non-preferred sequence is in the region of 0.2–0.5 kcal/mol. Thus, for an optimal site, bendability would contribute about 5 kcal/mol to the binding energy. For any protein–DNA complex the contribution of bending to complex formation will depend at least on the magnitude of curvature (or the extent of deformation) and the length of the curved DNA segment. Indeed it is notable that in both the CAP–DNA complex and the 434 operator–repressor complex, alteration of the sequence at a single bending site can change the binding affinity by 10–100-fold, that is equivalent to a change in binding energy of 1.5–3.0 kcal/mol. The difference between this value and that obtained for the core nucleosome is substantial and may reflect either differences in the methodology of the measurements or, alternatively, greater structural constraints in the short binding sites associated with CAP and with the 434 repressor.

An important feature of the average sequence organization of nucleosomal DNA is that the resultant bendability is anisotropic; that is, bending occurs only in a preferred direction. An everyday analogy of such a property is a bicycle chain, which is constructed to bend easily around the sprockets but cannot be bent at right angles to this direction without the application of considerable force. There are, however, certain DNA sequences which do not exhibit this sequence organization but which can nevertheless be incorporated into core nucleosome particles.

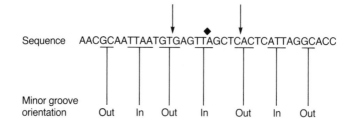

Figure 2.10 The sequence of the CAP binding site in the *lac* regulatory region showing the relation of the DNA sequence to the rotational orientation of the minor groove.

Such sequences, of which poly(dAT).(dAT), poly(dGT).(dAC) and poly(dAG).(dCT) are good examples, can accommodate the deformations associated with both a widening and a narrowing of the minor groove and are thus isotropically bendable or flexible. In nucleosomal DNA, as well as in promoter DNA, there is a tendency for such sequences to occur in particular regions, suggesting that the DNA in these may be deformed to an extent that is not simply accommodated by the short sequences normally associated with bendability. Where such extreme distortions ('kinks') occur in protein-bound DNA they are detectable by their reactivity with freely diffusible singlet oxygen. This reagent reacts with exposed bases, forming an adduct, which then sensitizes the adjacent sugar–phosphate backbone to alkaline hydrolysis. In double stranded DNA reactivity is highly dependent on base accessibility, which in turn is sensitive to, for example, the magnitude of curvature. The property of deformability, as with bendability, is sequence dependent and is correlated with a rapid rate of base pair opening.

2.4 DNA CONFIGURATION AND SEQUENCE PERIODICITY

The sequence preferences for DNA bendability are conserved in different nucleoprotein complexes and thus must reflect the intrinsic physicochemical properties of the DNA itself rather than the idiosyncrasies of a particular protein. The periodicity of these sequence preferences[*] reflects the local twist of the DNA on the surface of the protein, that is, if the DNA is uniformly bent, the number of base pairs between successive outward facing minor grooves. In the same way this periodicity defines the energetically preferred global configuration of the DNA molecule. For example, if the DNA is to be bent as part of a plane circle, then the local twist (defined as 360°/sequence periodicity) should be the same as the intrinsic twist of the DNA double helix. If, however, the average sequence periodicity differs from the intrinsic twist then the energetically preferred configuration of the molecule will be superhelical. The precise form of this superhelix depends crucially on the nature of the surface on which it is wrapped. The simplest case to consider is that of a cylinder, which is considered to be a reasonable approximation to the nucleosomal superhelix. In such a case, if the local twist is greater than the intrinsic twist then the sense of the superhelix will be left-handed (Figure 1.11). In the nucleosome both conditions are met, and thus the sense of the wrapping in the particle is compatible with the energetically preferred configuration. Conversely, if the local twist is less than the intrinsic twist, then the sense of the superhelix will be right-handed. When, however, the shape of the

[*]Note that sequence periodicity is measured in bp and twist is an angle between adjacent base pairs. As the twist increases so the number of bp per turn increases.

surface differs from a cylinder, which will be the usual case, such simple rules do not necessarily apply. For example, when the local twist is a little less than the intrinsic twist the DNA can still be wrapped as a left-handed superhelix, provided that the overall surface is barrel shaped. This is because when DNA is wound on a barrel the direction of curvature of the molecule relative to the superhelical axis changes with successive turns of the superhelix. Similarly, when the local twist is a little more than the intrinsic twist the DNA can be wrapped as a right-handed superhelix and satisfy the bending preferences of the DNA , provided that the surface is in the form of a capstan (Figure 1.13).

Although these relationships between local and intrinsic twist may seem at first sight somewhat abstruse it seems probable that they may be crucial to the understanding of the generation of local torsional stress within a nucleoprotein complex. In essence, if the DNA is forced by the protein into a configuration, or topology, which is incompatible with its defined bending preference, a local torsion could be generated which would result in local distortions of DNA conformation, such as kinking or base unstacking, leading to strand separation. Such a result could be achieved either by forcing a metastable initial wrapping under super-helical strain within the DNA or alternatively by a change in the shape of the wrapping surface following the initial binding of the DNA. Such a change would require substantial conformational movements in the binding protein(s), as are indeed observed in some processes involving the enzymatic manipulation of DNA, for example, in the transition from the closed to the open RNA polymerase–promoter complex (Chapter 4).

2.5 THE ESTABLISHMENT OF DNA ARCHITECTURE

Whenever a DNA molecule is bent the resultant DNA configuration reflects and requires local distortions in the double helix at the level of individual dinucleotide steps. In principle, any peptide structure which preferentially binds to any of the conformationally distinct distortions associated with bending could stabilize a bend in DNA and thereby define the local configuration of the DNA molecule. In chromatin, many different sequences must be accommodated in the overall configuration of the DNA molecule and it would be expected that stabilization of a particular configuration would not need to involve direct bonding of the peptide to individual bases. In eukaryotic histones, and certain other chromosomal proteins, a number of short peptide motifs have been iden-tified that preferentially bind to DNA sequences with unique conforma-tional characteristics.

These motifs fall into two classes, those that would be expected to assume some form of β-structure and those that are potentially α-helical in character (Table 2.1). Of the former the SPKK(SPRK) motif is repeated many times in the N and C termini of histone H1 and the N terminus of histone H2B from sea urchin sperm, in which the chromatin is very highly condensed. A dimer of this sequence, SPRKSPRK, by itself select- ively binds to the minor groove of DNA within short oligo(dA).(dT) tracts, a property which is consistent with the preferential recognition of the distinctly narrow minor groove characteristic of this DNA sequence. A second class of peptide with this property contains the GR(P) motif and is found in certain other chromosomal proteins, notably mammalian HMG-I and *Drosophila* D1. These proteins contain multiple repeats of sequences of the type PRKRGRPRK and again bind to the minor groove of DNA at A-T rich (although not necessarily oligo(dA).(dT)) sites. Although normally found in relatively abundant chromosomal proteins both the SPKK and GR(P) motifs also occur in DNA binding regulatory proteins. For example, the yeast protein SWI5, whose primary binding specificity is determined by three zinc fingers, contains both such motifs in a short peptide sequence, KRSPRKRGRPRK immediately adjacent to the zinc fingers. This sequence could act both as a nuclear localization signal and also provide additional binding energy resulting from a preferential interaction with an oligo(dA).(dT) tract within the SWI5 binding site. A closely related repeated sequence, GRKPG, is found in another yeast protein, datin, that exhibits a strong preference for longer (dA).(dT) tracts, which in principle would be more conformationally constrained than short tracts.

The α-helical motifs are again found largely in histones and other abundant chromosomal proteins. The best characterized of these is the AK helix which contains predominantly, and in some examples ex- clusively, the three amino acids alanine, lysine and proline. In such structures the lysine residues provide the positive charge to form electrostatic bonds to the sugar–phosphate backbone while alanine and proline function respectively to stabilize and to introduce kinks into the α-helix. Although AK helices are widely distributed in DNA binding proteins, both the frequency of occurrence of proline residues and the disposition of the lysines with respect to the helix are variable. These variations suggest that the AK helix is a versatile DNA binding motif in which a fundamental structure can be adapted to bind in different modes. The simplest of the naturally occurring AK structures found is the *Pseudomonas* regulatory protein AlgR3. This protein contains a region of about 170 amino acids, occupied principally by a tetra-residue repeat KPAA, among which are interspersed sequences of 9 or 5 residues. In a sequence of the form KPAAKPAA a kink of some 30–40° will be intro- duced by the second proline such that each KPAA repeat will form 1.1 α- helical turns (Figure 2.11). Because the repeat length of 4 residues is greater than the α-helical repeat of 3.6 residues per turn the helical axis

Table 2.1 Basic protein motifs for DNA binding

Motif/protein name	Sequence/structure[a,b]	DNA binding
(a) Proline-rich		
SPKK	β-turn?	A-T rich ≥ 5 bp
Histone H1 termini	} [ST]P[RK][KR] [> 6]	A-T rich DNA?
Sea urchin sperm H2B		
HBV coat protein	SPRRR(+) [3]	?
Mouse *c-myc*	SPRS, SPAR, etc.	?
Nucleolin	TP[GA]KKXX [6]	?
Human son 3	TPSRRSR [3]	?
Yeast SWI5	SPRKRGRPRK	?
GR(P)	G[RK]P; not SP or TP	
HMG-I and Y	[KR]P[RK}GR(K)P [2]	A-T rich ≥ 6 bp
Drosophila D1	(+)GRP(+) [10]	A-T rich DNA
Yeast datin	GRKPG [3]	A-T rich ≥ 8 bp
Yeast SWI5	SPRKRGRPRK	?
KPK	[KR]P[KR]	A-T rich DNA?
Histone H1 termini	KPK	?
Homeodomain	KRPR	A-T rich minor groove
434 repressor	KRPR	A-T rich minor groove
Others		
HU class	Basic β-ribbon arm	Minor groove Variable sequence
(b) α-Helical		
Decapeptide	Amphipathic helix	
Consensus for	P-VRKSLRKG	Nucleosome dyad?
H1, H4, SAP, CRP		
Histone H1	Variable	A-T rich DNA?
SAP	Variable	Nucleosome dyad?
AK	AAKK; AKKA, etc. α-helix	?
Histone H1 termini	AK-rich, variable	?
	Proline-free helix	dsDNA
AKP	Proline-kinked helix?	?
Histone H1 termini	e.g. P(KAAK)$_n$P	?
AlgR3	e.g. (KPAA)$_n$	?
Others		
RecA	24 residue helix	ssDNA
HMG 1 and 2	} Putative helical region	A-T rich DNA
hUBF (4 HMG 1 boxes)		G-C rich DNA

of a sequence containing many KPAA repeats will follow a right-handed superhelix with 10–11 repeats per superhelical turn with a radius of curvature of 23–27 Å. In such a structure the ε-amino groups of the lysine residues will be separated by 5–6 Å, that is by approximately the same distance as adjacent phosphate groups in a sugar–phosphate backbone of DNA. The geometry of this protein structure is thus ideally suited to track around the DNA double helix following a backbone. The interruptions of 5 or 9 amino acids between the regular repeating units would alter the rotational orientation of the lysine residues and consequently switch their contacts across a DNA groove to the other sugar–phosphate backbone.

A second example of an AK helix is found in the C-terminal tail of the H1 histone from the sperm of the sea urchin *Lyotechnicus pictus*. This domain consists of two sections, a 57-residue stretch devoid of proline residues and a 64-residue region containing, on average, a proline residue every 7.4 amino acids. In the latter region the repeat is not KPAA but KPA(A)KKAKK, with the lysine residues again restricted to one face of the kinked α-helix (Figure 2.12). However, in contrast to the AlgR3 structure, adjacent lysine residues are splayed out with a separation of some 11–18 Å between ε-amino groups, depending on the conformation of the lysine side chains. In the proline rich region this AK structure would have a radius of curvature of some 50–60 Å – that is, insufficient to track around a DNA double helix but comparable to the superhelical radius of nucleosomal wrapped DNA. Such a structure could thus interact with two adjacent turns of a DNA superhelix, a mode of interaction which would be fully compatible with the biological role of this region of H1 in promoting the condensation of the condensed 100 nm chromatin fibre. A third example is the *Drosophila* chromosomal protein HMG-N, in which a stretch

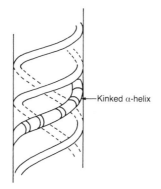

Figure 2.11 How an α-helix could wrap around DNA as a right-handed superhelix.

of 18 amino acid residues comprising five α-helical turns is arranged so that all the lysine residues are contained in a segment of 180°, while the alanine residues are found in an adjacent 90° segment.

The AK motif is an example of an amphipathic α-helix with positive charges concentrated on one face. Such a structure has the potential to bind to any regions of negative charge, whether they be in protein or nucleic acid. Interactions with DNA would be expected to be restricted primarily to the sugar–phosphate backbone and would not, by themselves, impart sequence selectivity. However, the overall configuration of the polypeptide chain has the potential to stabilize particular configurations of the DNA double helix and thereby facilitate the determination of the structure of eukaryotic and prokaryotic chromatin.

In addition to the ability of individual protein motifs to bind selectively to particular DNA structures, the primary function of certain proteins is to impart a particular global configuration to a DNA molecule. A notable example of such a protein is the *E. coli* integration host

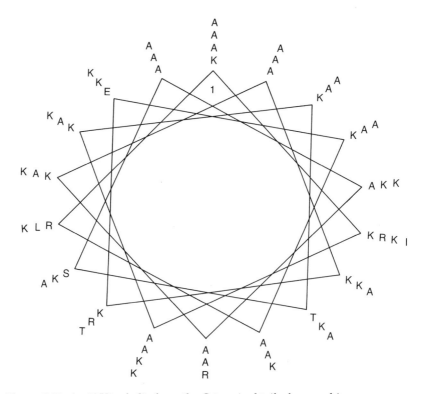

Figure 2.12 An 'AK' α-helix from the C-terminal tail of sea urchin sperm histone H1. Note that regions of positive charge (K) are essentially restricted to two faces of the helix.

factor (IHF) whose principle role is to bend DNA at specific sites. IHF is a heterodimer, closely related to the bacterial histone-like protein HU, which is believed to bind to the minor groove of DNA through a looped β-sheet structure (Figure 2.13). Unlike HU, however, IHF binds to a specific DNA sequence. When bound to such a site the heterodimer protects three double helical turns of DNA which are bent through 140°, that is, a curvature very similar to that in the nucleosome core particle. Genetic evidence shows that IHF is a necessary component for many processes requiring the assembly of large nucleoprotein complexes, notably recombination, DNA replication and transcription.

One such large nucleoprotein complex is the intasome. This complex mediates the integration of phage λ DNA into the host chromosome and contains several molecules of two DNA-binding proteins, the phage-coded integrase and IHF, bound to *attP*, the genetically defined attachment site on the phage DNA. In this complex the DNA is wrapped as a left-handed supercoil. The bivalent Int protein can bind simultaneously to two different segments of *attP* which are separated by a loop. This loop contains a binding site for IHF which bends the DNA and thus defines the trajectory of the loop. For this trajectory to be correctly defined in the intasome, the bends introduced by the three component IHF molecules need to be correctly phased relative to each other. If the sole function of IHF in the intasome were to act as an architectural element to specify the path of the DNA it should, in principle, be possible to overcome the requirement for IHF by substituting for the IHF binding site any other DNA element that has the ability to confer the appropriate curvature on the DNA. In just such an experiment, Goodman and Nash replaced one IHF binding site in *attP* by a site with high affinity for CAP, a protein which can bend the DNA to an extent comparable to that of IHF. In such constructs, significant recombination function was observed only in the presence of CAP and its activator, cAMP, suggesting that only the bending of DNA, and not additional heterologous protein–protein contacts, is necessary for the assembly of a functional intasome. This conclusion was confirmed by the finding that an intrinsically bent DNA sequence, with an estimated bend of approximately 120°, can also substitute for this same IHF binding site, provided that the bend is correctly phased and is sufficiently curved. However, the magnitude of curvature in intrinsically bent DNA is about 20° per double helical turn, that is, about half the value of the protein induced bending in the IHF and CAP complexes. Consequently, in order to bend the DNA through the same angle, twice as many double helical turns are required for intrinsically bent DNA as for the particular protein induced bends. This observation establishes that the function of the intasome requires that the bends in the wrapped DNA be in the correct direction and of the correct extent, but that the actual length of DNA included in the complex

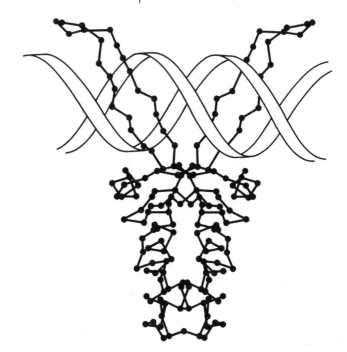

Figure 2.13 The structure of HU-protein dimer from *Bacillus stearothermophilus*. The binding of this protein in the minor groove of DNA is mediated by the loops of anti-parallel β-strands.

is not precisely specified.

The intasome is the best characterized example of a protein–DNA complex in which IHF functions as an architectural element. Other notable examples of such complexes include the complex established at the *E. coli* origin of DNA replication, *oriC*, with the DNA protein and transcription initiation complexes. However, IHF is only one example of a protein which stabilizes bends in DNA. Other such proteins include the *E. coli* FIS protein, involved in certain recombinogenic events and the regulation of transcription.

REFERENCES

Churchill, M.E.A. and Suzuki, M. (1989) 'SPKK' motifs prefer to bind to DNA at A/T-rich sites. *EMBO Journal*, **8**, 4189–95.
De Vargas, L.M., Kim, S. and Landy, A. (1989) DNA looping generated by the DNA bending protein IHF and the two domains of lambda integrase. *Science*, **244**, 1457–61.
Drew, H.R. and Travers, A.A. (1985) DNA bending and its relation to nucleosome positioning. *Journal of Molecular Biology*, **186**, 773–90.

Echols, H. (1986) Multiple DNA–protein interactions governing high precision DNA transactions. *Science*, **233**, 1050–66.

Gartenberg, M.C. and Crothers, D.M. (1988) Sequence determinants of CAP-induced bending and protein binding affinity. *Nature*, **333**, 824–9.

Goodman, S.D. and Nash, H.A. (1989) Functional replacement of a protein-induced bend in a DNA recombination site. *Nature*, **341**, 251–4.

Richmond, T.J., Finch, J.T., Rushton, B. *et al.* (1984) Structure of the nucleosome core particle at 7 Å resolution. *Nature*, **311**, 532–7.

Satchwell, S.C., Drew, H.R. and Travers, A.A. (1986) Sequence periodicities in nucleosome core DNA. *Journal of Molecular Biology*, **191**, 659–75.

Schultz, S.C. *et al.* (1991) Crystal structure of a CAP–DNA complex: the DNA is bent by 90°. *Science*, **253**, 1001–7.

Suck, D. Lahm, A. and Oefner, C. (1988) Structure refined to 2 Å of a nicked DNA octanucleotide complex with DNase I. *Nature*, **332**, 464–8.

3

DNA–protein interactions: sequence specific recognition

3.1 GENERAL PRINCIPLES OF SEQUENCE SPECIFIC RECOGNITION

The binding of a protein to a specific DNA sequence is largely dependent on two types of interaction. The principal basis for sequence selectivity is direct contact between the polypeptide chain and the exposed edges of the base pairs, primarily in the major groove of B-DNA. These contacts may involve either hydrogen bonds or van der Waals interactions. Molecules which are tightly and rigidly bound to a protein and are thus integral components of the macromolecular structure may also participate in these interactions and so provide binding specificity to the protein by proxy. These direct interactions are supplemented by the sequence dependent bendability or deformability of DNA, which limits the energetically favourable conformations of a particular binding site and thereby imposes additional sequence dependent constraints on the binding affinity. It should, however, be emphasized that, relative to direct contacts to the base pairs, this is a second order effect which modulates the affinity for the whole binding site.

The binding energy available from direct interactions with the base pairs, although significant, is not in general sufficient by itself to allow the formation of a stably bound complex for binding sites of average length (6–15 bp). The required additional binding energy may be provided by direct electrostatic interactions between basic amino acid residues and the negatively charged sugar–phosphate backbone. The spatial constraints imposed by this type of interaction may also serve to constrain the configuration of the DNA when bound to protein. An important additional characteristic of these electrostatic bonds is that they enable the protein to bind to any region of DNA with the appropriate backbone configuration. That is, the protein possesses the capability for sequence independent interactions with the DNA double helix. It is the difference between the binding energies for the sequence dependent and sequence independent components of the interaction that is the measure of the sequence selectivity of a DNA-binding protein.

3.2 STRUCTURAL MOTIFS FOR SEQUENCE SPECIFIC BINDING

The recognition of a specific DNA sequence by a polypeptide chain can be achieved by a diverse but limited range of structures (Table 3.1). Structural, biochemical and molecular genetic studies of protein–DNA interactions have identified a small number of defined protein structural elements that bind to specific DNA sequences. Those of the most frequent occurrence in characterized proteins include the helix–turn–helix motif characteristic of many eukaryotic regulatory proteins, the zinc finger motif found in DNA and RNA binding proteins in most organisms, an α-helical structure found in the *Eco*RI restriction enzyme and antiparallel β-strands in the *E. coli met* repressor. In all these examples the recognition motif fits into the major groove of DNA and must therefore protrude from the surface of the protein. Nevertheless, for one protein, *E. coli* integration host factor, it has been suggested that an antiparallel β-sheet might confer sequence selectivity by binding in the minor groove. In addition to motifs for which structural information is available other DNA binding regions of proteins have been identified by sequence homology. These include the proteins of the helix–loop–helix class and also those which contain a leucine zipper dimerization domain.

3.2.1 The helix–turn–helix motif

The helix–turn–helix motif was first identified as a conserved sequence element in the repressors encoded by the lambdoid phages of *E. coli* and *Salmonella typhimurium*. Subsequently, this element has been found in a large variety of DNA-binding proteins, both in prokaryotes and in eukaryotes; it shows a very high degree of structural conservation. This DNA-binding domain consists of two short α-helices which are separated by a glycine residue. This amino acid, in concert with its neighbours, acts as a flexible hinge allowing the polypeptide chain to bend between the two helices so that they can make hydrophobic contacts with each other (Figure 3.1). These contacts preserve the relative orientation of the two helices and result in the formation of a compact tertiary structural domain. Although the structure of this domain is highly conserved the orientation of the motif to its DNA binding site is quite variable. This feature suggests that the motif is a structure which possesses a rigidity which is necessary for precise sequence recognition. Such a property could also be utilized for other protein functions and indeed, there is at least one example of a DNA-binding protein in which this motif does not constitute the DNA-binding domain.

Table 3.1 Protein structures for DNA binding

α-helices

(a) AK(P)motif	not sequence selective presumed contacts with sugar-phosphate backbone	Histone H1 AlgR3
(b) helix-turn-helix	sequence-selective major groove contact orientation in groove variable	bacterial repressors /activators
(c) homeodomain	sequence-selective major groove contacts with a few minor groove contacts	homeobox proteins
(d) Anfingers (TFIIIA type)	sequence selective major groove contacts bind DNA and RNA	TFIIIA
(e) Zn fingers (steroid receptor type)	sequence selective major groove contacts	steroid receptors
(f) Zn fingers (GAL4 type)	sequence selective major groove contacts	GAL4
(g) Bzip	sequence-selective major groove contacts	GCN4; *fos*; *jun*
(h) parallel α-helices	sequence-selective major groove contacts (DNA or RNA)	EcoR1 restriction enzyme Glutaminyl tRNA synthetase (RNA)

β-Structures

(a) anti-parallel strands	not sequence selective sequence selective minor groove contacts major groove contacts	*E. coli* HU protein *E. coli* IHF; TFIID *E. coli* MetJ protein; P22 *arc* and *mnt* repressors
(b) GR(P) and SPKK motifs	structure selective minor groove interactions	Histone H1; datin; nucleolin

Other structures (undetermined)

(a) helix-loop-helix	α-helical dimerization domain (recognition domain assumed to be α-helical)	MyoD1; scute
(b) HMG-box	sequence/structure selective minor groove contacts	HMG-1; UBF-1; Lef-1
(c) Y-box		FRY-1

This list is not exhaustive but emphasizes the variety of motifs that bind to DNA and the preponderance of major groove interactions for sequence-specific recognition.

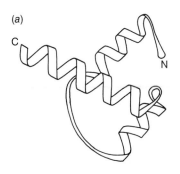

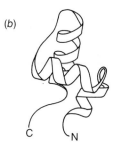

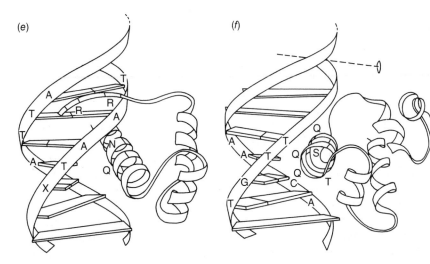

Figure 3.1 Variations on a theme. Structural variations in DNA-binding domains incorporating a helix-turn-helix motif: (a) homeodomain; (b) the *lac* repressor headpiece; (c) the helix-turn-helix motif in the C$_I$ repressors of phages λ and 434; (d) the CAP DNA-binding domain; (e) interaction of the homeodomain with DNA. Note the contacts of the N-terminal arm in the minor groove; (f) the 434 repressor binding domain bound to an operator half-site.

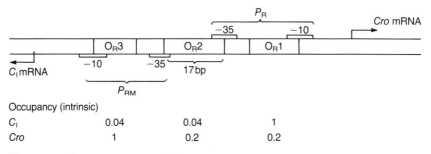

Figure 3.2 The organization of the right-hand operator of λ phage. The relative affinities of the C_I and *cro* repressor for each of the individual operator sites is shown.

C_I and cro repressors

The lambdoid phages of *E. coli* and *Salmonella typhimurium* encode two repressor proteins C_I and *cro*. Both of these repressors bind to a set of six similar, but non-identical, sites or operators in the phage genome. These sites are grouped in sets of three on each side of the structural gene for the C_I protein. One set, including the operators O_R1, O_R2 and O_R3, overlaps two promoters: P_R, which directs the synthesis of a set of genes expressed immediately after phage infection or induction, and P_{RM} which directs the production of the mRNA for the C_I repressor itself. The second set of operators, O_L1, O_L2 and O_L3, lies on the other side of the C_I gene and overlaps the P_L promoter which directs the synthesis of another set of early mRNA species (Figure 3.2). The affinities of the C_I repressor for these six sites differ as also do those of the *cro* repressor. It is this differential affinity which creates a regulatory switch determining the difference between lysogeny and lytic growth.

The structures of these two repressor proteins differ substantially. Whereas the *cro* protein of phage λ contains 56 amino acids and exists in solution principally as a dimer, the C_I repressor contains 236 amino acids and can exist as a monomer, a dimer or a tetramer in solution. Both proteins bind to the individual operator sites as dimers. The greater complexity of the C_I repressor is also apparent on gently proteolysis. When this protein is treated with low concentrations of trypsin it is cleaved into two domains: an N-terminal domain of some 90 amino acids which contains the helix–turn–helix motif and binds specifically to the operator sites, and a C-terminal domain which does not interact with DNA. Between these two regions is a flexible hinge. This two-domain structure is also found in other DNA-binding proteins, notably the C_I repressors of other lambdoid phages, the *lac* repressor and the LexA protein which represses expression of the SOS genes involved in DNA repair.

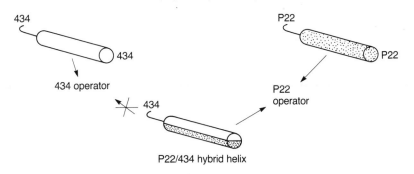

Figure 3.3 Helix swap experiment to alter the specificity of repressor binding.

How is the selectivity of binding of the helix–turn–helix motif established? The evidence that this structure is the principal determinant of sequence recognition in the lambdoid repressors is based both on genetic analysis and on 'helix swap' experiments. Although these phages are closely related, the repressor proteins from different phages recognize different sequences. However, because the three-dimensional structure of these different proteins is very similar, it is possible, in principle, to exchange, by genetic engineering, the helix–turn–helix structure of one protein with that of its relative. In one such experiment the putative DNA-binding domain of the *cro* repressor of phage P22 was substituted for the corresponding domain in the *cro* protein of phage 434. This construction involved the substitution of only those amino acids in the recognition helix that were on the face that contacted the base pairs. In this way the interactions of the opposite face with the rest of the domain were retained. The result of this experiment was a switch of sequence specific recognition, from the sequence normally recognized by the 434 *cro* protein to that recognized by P22 *cro* protein (Figure 3.3). This experiment directly demonstrates that, in these proteins, the helix–turn–helix domain determines sequence recognition. For a second protein, the λ C_I repressor, genetic evidence again shows that the helix–turn–helix motif is required for DNA recognition. By selecting for mutant C_I proteins that fail to bind to the operator site and are not themselves grossly disrupted in tertiary structure it was found that the majority of mutations affecting DNA-binding are localized to the helix–turn–helix region.

For the second λ encoded repressor, the C_I protein, genetic evidence again shows that the helix–turn–helix motif is required for DNA binding. By selecting for mutant C_I proteins that fail to bind to the operator site and identifying the mutations so obtained, it was found that the majority of strong mutations affecting DNA binding were localized to the helix–turn–helix region. Although genetic experiments identify the helix–turn–helix motif as the principal structural element of the C_I repressor required for direct interactions with the operator sites, additional contacts between the protein and the DNA also contribute significantly to the affinity of the protein for the operator. These contacts are made by a short stretch of amino acids at the N-terminus. Removal of the first three N-terminal amino acids by proteolysis reduces the binding constant for the N-terminal domain by about 50-fold. Analysis of the crystal structure of the whole N-terminal domain by itself shows that the N-terminus forms a free tail which has the potential to contact the DNA at the operator dyad on the opposite face of the DNA to the major groove contacts made by the helix–turn–helix motif.

The precise nature of the interactions of the helix–turn–helix motifs of the lambdoid repressors with their cognate operator sites has been determined in three cases from crystal structures of cocrystals of the N-terminal domains of the λ and 434 C_I repressors and the 434 *cro* repressor with their binding sites. The principal elements of sequence specific recognition in these structures are hydrogen bonds and van der Waals contacts between amino acid residues in the second helix and base pairs. In addition to these base specific contacts both of the helices contact the backbone phosphates. Interestingly, in the case of the 434 C_I repressor three of these contacts are made between —NH— groups in the peptide backbone while only one phosphate contact involves a negatively charged amino acid side chain.

Although the general nature of the interactions between the helix–turn–helix motif itself and the major groove of DNA is similar for all these repressors the overall geometry of their complexes differs. Whereas in the λ C_I repressor–operator complex the helical axis of the DNA is essentially straight and the DNA undistorted, in that involving the 434 repressor the operator DNA is distorted with an overall bend with an average radius of curvature of 65 Å. As in the nucleosome, bending is not smooth but is concentrated at two sites symmetrically disposed 2–3 base steps away from the central region of the operator. The most striking feature of the structure is a narrowing of the minor groove at the centre to 8.8 Å compared with an average of 11.5 Å for B-DNA. This narrowing is accompanied by a significant overwinding of the base steps in this region. The crucial importance of this central region of the operator for the interaction with the protein was demonstrated by the observation that the introduction of a nick on one strand of the operator DNA at its central phosphodiester bond increased the affinity of the operator for the repressor by approximately 5-fold. Such a nick would increase the flexibility of the sugar–phosphate backbone in this region and so facilitate the correct disposition of each of the monomer units of the protein with respect to each other.

Within the central region of the 434 operator there are no direct amino acid base pair contacts, suggesting that within this region the local conformation of the DNA is important. Consistent with this view, this region is mutationally sensitive, such that mutations to sequences which cannot readily adopt a narrow minor groove (in general, GC rich sequences) result in a significant decrease in affinity of the operator for the repressor, while mutations to AT rich sequences (in particular to a short (dA).(dT) run which readily adopts such a conformation), have the opposite effect. In the 434 *cro* repressor–operator complex the overall bend in the DNA is of comparable magnitude to that in the C_I repressor complex, yet the local conformation of the DNA in the central region differs subtly in respect to minor groove width in the two complexes. In the *cro* complex, a slightly wider minor groove in the central region, although still narrow relative to an average width, is reflected in a reduced response to sequence change when compared with the C_I complex.

Catabolite activator protein (CAP)

The catabolite activator protein from *E. coli* is responsible for activating the transcription of many operons involved in the uptake and catabolism of different sugars. In addition, it functions as a repressor of the transcription of its own gene. When the intracellular concentration of glucose falls in *E. coli* the level of cAMP rises substantially. This nucleotide binds to CAP, thereby increasing the affinity of the protein for its binding sites. CAP is a homodimer of a 22 500 Da polypeptide which contains a helix–turn–helix motif. In the presence of cAMP this dimer protects approximately three double helical turns of DNA. Two features of this interaction are of particular interest: the nature of the interaction of the DNA-binding domain with the target site and the nature of the DNA bending induced by the protein. The target site contains an interrupted inverted repeat with a highly conserved TGTG/CACA sequence located half a turn away from the centre of symmetry (Figure 2.10). This sequence is contacted by the helix–turn–helix motif in the major groove, into which the recognition helix protrudes more deeply than in the repressor–operator complexes; it is parallel to the base pairs. In the crystal structure the consequence of this interaction is to close the major groove and enormously widen the opposing minor groove. Such changes in groove width are brought about by a 40° kink produced by base pair roll at a TG base step. In this situation the bending of the double helical axis is essentially localized to a single step; in other words the bending is discontinuous (Figure 3.4).

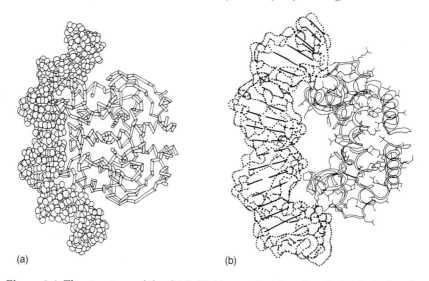

(a)   (b)

Figure 3.4 The structure of the CAP–DNA complex showing the kinks induced by CAP at TG/CA steps in the binding site.

lac *repressor protein*

The *lac* repressor protein is required in addition to CAP for the regulation of the *lac* operon in *E. coli*. In the absence of galactose the repressor binds to an operator site centred 10.5 bp upstream of the transcription startpoint and blocks initiation of RNA synthesis by RNA polymerase. In the presence of galactose the repressor binds the metabolite allolactose which reduces the affinity of the repressor for the operator, allowing transcription to proceed.

The *lac* repressor protein is a tetramer of four identical subunits of 38 000 Da, each of which contains a helix–turn–helix motif as the principal DNA-binding region within the N-terminal domain of the polypeptide. Although the structure of this domain has not been solved by crystallographic methods, nuclear magnetic resonance spectroscopy has identified contacts between the domain and the operator site. The principal conclusion from these studies is that the second helix of the helix–turn–helix motif is placed in the major groove of DNA but in the opposite orientation to that assumed by the corresponding helix in the lambdoid repressor proteins and CAP; that is, the amino end of the 'recognition' helix interacts with base pairs closer to the dyad axis of the operator than do side chains in the centre of the helix (Figure 3.5). It should be noted that the binding of the *lac* repressor to a single operator site is mediated by only two of the component subunits, leaving the two remaining subunits free to interact with another operator site.

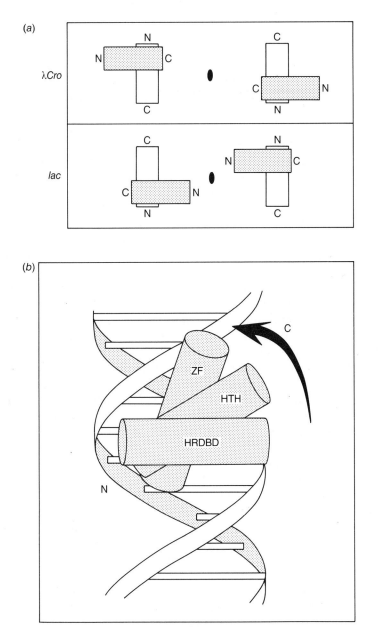

Figure 3.5 (a) The orientation of the helix-turn-helix motif in the major groove relative to the mid-point of the operator in the λ *cro* repressor and the *lac* repressor. (b) The orientation of the recognition helix in the major groove of DNA for three classes of DNA-binding proteins : zinc finger (ZF) of the TFIIIA type, helix-turn-helix (HTH) and steroid receptors (HRDBD).

Tryptophan repressor

In the previous example, the selectivity of the protein for a particular DNA sequence depends both on direct contacts between the protein and the bases and also on local conformational constraints on the DNA structure. However, in the case of the tryptophan repressor, substantial selectivity can be achieved in the absence of such direct contacts. In the presence of the ligand L-tryptophan this repressor binds to an operator site within the promoter region of the *trp* operon and blocks transcription initiation. This operon encodes the enzymes required for the biosynthesis of tryptophan which, on binding to the Trp repressor, completes a negative feedback loop.

The Trp repressor has been crystallized as a complex with a symmetrized 18 bp operator, TGTACTAGTTAACTAGTAC. In this sequence the base pairs whose function *in vivo* is most sensitive to mutation are the two ACTAG sequences centred 4–5 bp from the dyad. In the cocrystal there are no direct hydrogen bonded or nonpolar contacts between the helix–turn–helix motif and the DNA that can account for the selectivity of the interaction. (The one direct contact between a base and an amino acid side chain lies outside the mutationally sensitive region.) Instead, direct hydrogen bonded contacts are made principally to the phosphate groups in the DNA backbone. These contacts serve to constrain the backbone in a defined configuration. However, three well ordered and directed water molecules lying in the major groove between the helix–turn–helix motif and the CTAG sequence make hydrogen bonded contacts with the base pairs in this sequence. Notably, recognition of the adenine residue immediately preceding the CTAG sequence by a water-mediated hydrogen bond is sequence specific (Figure 3.6). Mutations in the amino acids that bind to these water molecules result both in a reduced affinity, and, in some examples, an altered sequence selectivity of the repressor for the operator. This structure thus suggests that a degree of sequence selectivity could be achieved by water molecules forming an essential component of the recognition surface of the protein and thus mediating direct read-out of the operator sequence acting, as it were, by proxy for the protein.

Another feature of the cocrystal is the structure of the DNA which is dominated by the flexibility of the TpA steps. At each of the three central TpA dinucleotides there is a significant bend in the helical axis of the DNA. The bend at the sequence dyad directs curvature away from the protein, whereas the two flanking TpA steps in the CTAG sequence direct curvature towards the protein, a pattern of bending that is very similar to that observed in CAP. The flexibility of these TpA steps could in principle contribute to the formation of a stable DNA complex by

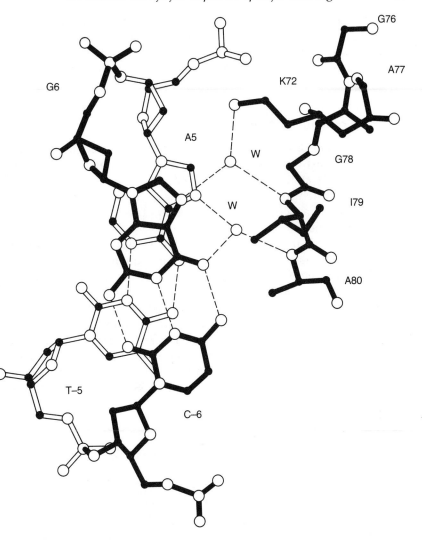

Figure 3.6 The position of constrained water molecules forming hydrogen bonds between the recognition helix of the Trp repressor and the conserved region of the *trp* operator.

providing a low energy barrier to any necessary structural rearrangement on binding of the protein.

It should be noted that the nature of the physiological binding site for a single Trp repressor dimer is controversial. In particular, evidence has been presented which has been interpreted to show that the binding site for a single dimer in the operator sequence is centred, not at the mid-point of the binding site utilized in the cocrystal, but instead at a site 4 bp to the right or to the left of this dyad. In other words, on this latter proposal each *trp* operator would bind two dimers. However, the most recent evidence strongly supports the notion that the natural binding site for the repressor dimer is the canonical operator.

3.2.2 Zinc-containing DNA-binding domains

A second class of protein motif for sequence specific recognition of DNA comprises at least three structurally distinct mini-domains which share the property of stabilization by a tetrahedrally co-ordinated zinc ion. The first of these structures to be recognized was the archetypal 'zinc finger' present in the transcription factor TFIIIA. This protein was initially purified from oocytes of the frog *Xenopus laevis* and functions as a positively acting transcription factor for the 5S rRNA genes, binding to a 50 bp region contained within the coding sequence for 5S RNA (the internal control region). In addition TFIIIA is found in large quantities in mature oocytes in the form of 7S complex with 5S RNA itself. The protein thus has the capability of binding to both DNA and RNA.

Proteolytic studies show that the TFIIIA protein, with a molecular weight of 40 kDa, consists of two functionally separable regions. The first, a 10 kDa domain at the carboxy-terminus, does not appear to be involved in DNA binding but is essential for efficient transcription, while the remaining 30 kDa domain binds both to 5S RNA and to the whole

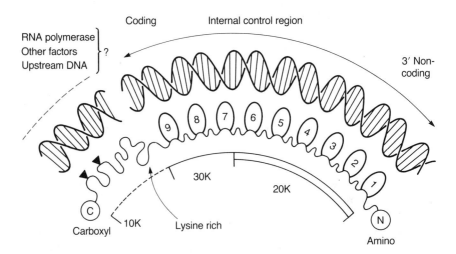

Figure 3.7 A schematic representation of the TFIIIA protein and its binding to the internal control region of the DNA of the 5S rRNA gene.